Two Manuals Combined into One Book

AF812691

# Special Forces Survival Evasion and Recovery

# And

# Rappelling

Department of Defense

Pentagon
Publishing

## QUICK REFERENCE CHECKLIST

### Decide to Survive!

**S** - **S**ize up the situation, surroundings, physical condition, equipment.
**U** - **U**se all your senses
**R** - **R**emember where you are.
**V** - **V**anquish fear and panic.
**I** - **I**mprovise and improve.
**V** - **V**alue living.
**A** - **A**ct like the natives.
**L** - **L**ive by your wits.

**1. Immediate Actions**

a. Assess immediate situation. ***THINK BEFORE YOU ACT!***

b. Take action to protect yourself from nuclear, biological, or chemical hazards (Chapter IX).

c. Seek a concealed site.

d. Assess medical condition; treat as necessary (Chapter V).

e. Sanitize uniform of potentially compromising information.

f. Sanitize area; hide equipment you are leaving.

g. Apply personal camouflage.

h. Move away from concealed site, zigzag pattern recommended.

i. Use terrain to advantage, communication, and concealment.

j. Find a hole-up site.

**2. Hole-Up-Site** (Chapter I)

a. Reassess situation; treat injuries, then inventory equipment.

b. Review plan of action; establish priorities (Chapter VI).

c. Determine current location.

d. Improve camouflage.

e. Focus thoughts on task(s) at hand.

f. Execute plan of action. Stay flexible!

***Recommend inclusion of this manual in the aviator's survival vest.***

**3. Concealment** (Chapter I)

a. Select a place of concealment providing—

(1) Adequate concealment, ground and air.

(2) Safe distance from enemy positions and lines of communications (LOC).

(3) Listening and observation points.

(4) Multiple avenues of escape.

(5) Protection from the environment.

(6) Possible communications/signaling opportunities.

b. Stay alert, maintain security.

c. Drink water.

**4. Movement** (Chapters I and II)

a. Travel slowly and deliberately.

b. ***DO NOT*** leave evidence of travel; use noise and light discipline.

c. Stay away from LOC.

d. Stop, look, listen, and smell; take appropriate action(s).

e. Move from one concealed area to another.

f. Use evasion movement techniques (Chapter I).

**5. Communications and Signaling** (Chapter III)

a. Communicate as directed in applicable plans/orders, particularly when considering transmitting ***in the blind***.

b. Be prepared to use communications and signaling devices on short notice.

c. Use of communications and signaling devices may compromise position.

**6. Recovery** (Chapter IV)

a. Select site(s) IAW criteria in theater recovery plans.

b. Ensure site is free of hazards; secure personal gear.

c. Select best area for communications and signaling devices.

d. Observe site for proximity to enemy activity and LOC.

e. Follow recovery force instructions.

FM 21-76-1
MCRP 3-02H
NWP 3-50.3
AFTTP(I) 3-2.26

FM 21-76-1 U.S. Army Training and Doctrine Command
Fort Monroe, Virginia

MCRP 3-02H Marine Corps Combat Development Command
Quantico, Virginia

NWP 3-50.3 Navy Warfare Development Command
Newport, Rhode Island

AFTTP(I) 3-2.26 Headquarters Air Force Doctrine Center
Maxwell Air Force Base, Alabama

29 JUNE 1999

# Survival, Evasion, and Recovery

## Multiservice Procedures for Survival, Evasion, and Recovery

**Note:** This *UNCLASSIFIED* publication is designed to provide Service members quick-reference survival, evasion, and recovery information. See Appendix B for the scope, purpose, application, implementation plan, and user information.

## TABLE OF CONTENTS

# THE CODE OF CONDUCT

## ARTICLE I

I am an American, fighting in the forces which guard my country and our way of life. I am prepared to give my life in their defense.

## ARTICLE II

I will never surrender of my own free will. If in command, I will never surrender the members of my command while they still have the means to resist.

## ARTICLE III

If I am captured, I will continue to resist by all means available. I will make every effort to escape and aid others to escape. I will accept neither parole nor special favors from the enemy.

## ARTICLE IV

If I become a prisoner of war, I will keep faith with my fellow prisoners. I will give no information or take part in any action which might be harmful to my comrades. If I am senior, I will take command. If not, I will obey the lawful orders of those appointed over me and will back them up in every way.

## ARTICLE V

When questioned, should I become a prisoner of war, I am required to give name, rank, service number and date of birth. I will evade answering further questions to the utmost of my ability. I will make no oral or written statements disloyal to my country and its allies or harmful to their cause.

## ARTICLE VI

I will never forget that I am an American, fighting for freedom, responsible for my actions, and dedicated to the principles which made my country free. I will trust in my God and in the United States of America.

## Chapter I

# EVASION

**1. Planning**

a. Review the quick reference checklist on the inside cover.

b. Guidelines for successful evasion include—

(1) Keeping a positive attitude.

(2) Using established procedures.

(3) Following your evasion plan of action.

(4) Being patient.

(5) Drinking water (***DO NOT*** eat food without water).

(6) Conserving strength for critical periods.

(7) Resting and sleeping as much as possible.

(8) Staying out of sight.

c. The following odors stand out and may give an evader away:

(1) Scented soaps and shampoos.

(2) Shaving cream, after-shave lotion, or other cosmetics.

(3) Insect repellent (camouflage stick is least scented).

(4) Gum and candy (smell is strong or sweet).

(5) Tobacco (odor is unmistakable).

d. Where to go (initiate evasion plan of action):

(1) Near a suitable area for recovery.

(2) Selected area for evasion.

(3) Neutral or friendly country or area.

(4) Designated area for recovery.

**2. Camouflage**

a. Basic principles:

(1) Disturb the area as little as possible.

(2) Avoid activity that reveals movement to the enemy.

(3) Apply personal camouflage.

b. Camouflage patterns **(Figure I-1)**:

(1) Blotch pattern.

(a) Temperate deciduous (leaf shedding) areas.

(b) Desert areas (barren).

(c) Snow (barren).

(2) Slash pattern.

(a) Coniferous areas (broad slashes).

(b) Jungle areas (broad slashes).
(c) Grass (narrow slashes).

(3) Combination. May use blotched and slash together.

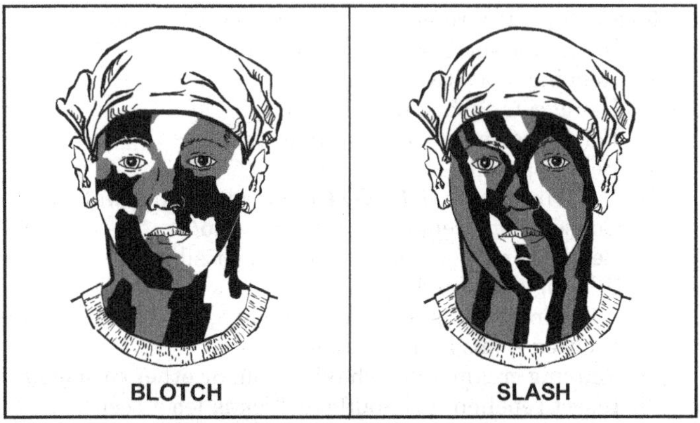


**Figure I-1. Camouflage Patterns**

c. Personal camouflage application follows:

(1) Face. Use dark colors on high spots and light colors on any remaining exposed areas. Use a hat, netting, or mask if available.

(2) Ears. The insides and the backs should have **2** colors to break up outlines.

(3) Head, neck, hands, and the under chin. Use scarf, collar, vegetation, netting, or coloration methods.

(4) Light colored hair. Give special attention to conceal with a scarf or mosquito head net.

d. Position and movement camouflage follows:

(1) Avoid unnecessary movement.

(2) Take advantage of natural concealment:

(a) Cut foliage fades and wilts, change regularly.
(b) Change camouflage depending on the surroundings.
(c) ***DO NOT*** select vegetation from same source.
(d) Use stains from grasses, berries, dirt, and charcoal.

(3) ***DO NOT*** over camouflage.

(4) Remember when using shadows, they shift with the sun.

(5) Never expose shiny objects (like a watch, glasses, or pens).

(6) Ensure watch alarms and hourly chimes are turned off.

(7) Remove unit patches, name tags, rank insignia, etc.

(8) Break up the outline of the body, ***"V"*** of crotch/armpits.

(9) Conduct observation from a prone and concealed position.

**3. Shelters**

a. Use camouflage and concealment.

b. Locate carefully—easy to remember acronym: ***BLISS***.

| |
|---|
| **B** - **B**lend<br>**L** - **L**ow silhouette<br>**I** - **I**rregular shape<br>S - Small<br>S - Secluded location |

(1) Choose an area—

(a) Least likely to be searched (drainages, rough terrain, etc.) and blends with the environment.

(b) With escape routes (***DO NOT*** corner yourself).

(c) With observable approaches.

(2) Locate entrances and exits in brush and along ridges, ditches, and rocks to keep from forming paths to site.

(3) Be wary of flash floods in ravines and canyons.

(4) Conceal with minimal to no preparation.

(5) Take the direction finding threat into account before transmitting from shelter.

(6) Ensure overhead concealment.

**4. Movement**

a. A moving object is easy to spot. If travel is necessary—

(1) Mask with natural cover **(Figure I-2)**.

(2) Use the military crest.

(3) Restrict to periods of low light, bad weather, wind, or reduced enemy activity.

Figure I-2. Ground Movement

(4) Avoid silhouetting **(Figure I-3)**.

(5) At irregular intervals—

(a) ***STOP*** at a point of concealment.

(b) ***LOOK*** for signs of human or animal activity (smoke, tracks, roads, troops, vehicles, aircraft, wire, buildings, etc.). Watch for trip wires or booby traps and avoid leaving evidence of travel. Peripheral vision is more effective for recognizing movement at night and twilight.

(c) ***LISTEN*** for vehicles, troops, aircraft, weapons, animals, etc.

(d) ***SMELL*** for vehicles, troops, animals, fires, etc.

**Figure I-3. Avoid Silhouetting**

(6) Employ noise discipline; check clothing and equipment for items that could make noise during movement and secure them.

b. Break up the human shape or recognizable lines.

c. Route selection requires detailed planning and special techniques (irregular route/zigzag) to camouflage evidence of travel.

d. Some techniques for concealing evidence of travel follows:

(1) Avoid disturbing the vegetation above knee level.

(2) ***DO NOT*** break branches, leaves, or grass.

(3) Use a walking stick to part vegetation and push it back to its original position.

(4) ***DO NOT*** grab small trees or brush. (This may scuff the bark or create movement that is easily spotted. In snow country, this creates a path of snowless vegetation revealing your route.)

(5) Pick firm footing (carefully place the foot lightly but squarely on the surface to avoid slipping). ***TRY NOT TO—***

(a) Overturn ground cover, rocks, and sticks.

(b) Scuff bark on logs and sticks.

(c) Make noise by breaking sticks. (Cloth wrapped around feet helps muffle this.)

(d) Mangle grass and bushes that normally spring back.

(6) Mask unavoidable tracks in soft footing by—

(a) Placing tracks in the shadows of vegetation, downed logs, and snowdrifts.

(b) Moving before and during precipitation allows tracks to fill in.

(c) Traveling during windy periods.

(d) Taking advantage of solid surfaces (logs, rocks, etc.) leaving less evidence of travel.

(e) Patting out tracks lightly to speed their breakdown or make them look old.

(7) Secure trash or loose equipment—hide or bury discarded items. (Trash or lost equipment identifies who lost it.)

(8) Concentrate on defeating the handler if pursued by dogs.

e. Penetrate obstacles as follows:

(1) Enter deep ditches feet first to avoid injury.

(2) Go around chain-link and wire fences. Go under fence if unavoidable, crossing at damaged areas. ***DO NOT*** touch fence; look for electrical insulators or security devices.

(3) Penetrate rail fences, passing under or between lower rails. If impractical, go over the top, presenting as low a silhouette as possible **(Figure I-4)**.

(4) Cross roads after observation from concealment to determine enemy activity. Cross at points offering concealment such as bushes, shadows, bend in road, etc. Cross in a manner leaving your footprints parallel (cross step sideways) to the road. **(Figure I-5)**

(5) Use same method of observation for railroad tracks that was used for roads. Next, align body parallel to tracks with face down, cross tracks using a semi-pushup motion. Repeat for the second track. **(Figure I-6)**.

Figure I-4.  Rail Fences

Figure I-5.  Road Crossing

Figure I-6. Railroad Tracks

**WARNING: If 3 rails exist, 1 may be electrified.**

## Chapter II

# NAVIGATION

| Assess the threat and apply appropriate evasion principles. |
|---|

**1. Stay or Move Considerations**

a. Stay with the vehicle/aircraft in a non-combat environment.

b. Leave only when—

(1) Dictated by the threat.

(2) Are certain of your location, have a known destination, and have the ability to get there.

(3) Can reach water, food, shelter, and/or help.

(4) Convinced rescue is not coming.

c. Consider the following if you decide to travel:

(1) Follow the briefed evasion plan.

(2) Determine which direction to travel and why.

(3) Decide what equipment to take, cache, or destroy.

d. Leave information at your starting point (in a non-combat environment) that includes—

(1) Destination.

(2) Route of travel.

(3) Personal condition.

(4) Supplies available.

e. Consider the following for maps (in a combat environment):

(1) ***DO NOT*** write on the map.

(2) ***DO NOT*** soil the map by touching the destination.

(3) ***DO NOT*** fold in a manner providing travel information.

Note: **These actions may compromise information if captured.**

**2. Navigation and Position Determination**

a. Determine your general location by—

(1) Developing a working knowledge of the operational area.

(a) Geographic checkpoints.

(b) Man-made checkpoints.

(c) Previous knowledge of operational area.

(2) Using the ***Rate x Time = Distance*** formula.

(3) Using information provided in the map legend.

(4) Using prominent landmarks.

(5) Visualizing map to determine position.

b. Determine cardinal directions (north, south, east, and west) by—

(1) Using compass.

**CAUTION:** The following methods are ***NOT*** highly accurate and give only general cardinal direction.

(2) Using stick and shadow method to determine a true north-south line **(Figure II-1)**.

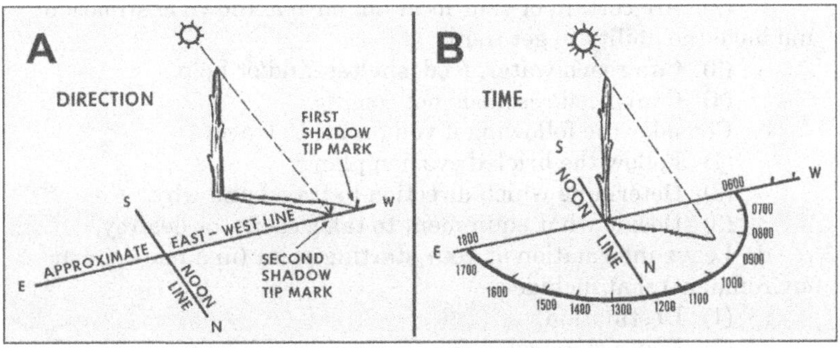


**Figure II-1. Stick and Shadow Method**

(3) Remembering the sunrise/moonrise is in the east and sunset/moonset is in the west.

(4) Using a wristwatch to determine general cardinal direction **(Figure II-2)**.

(a) Digital watches. Visualize a clock face on the watch.

(b) Northern Hemisphere. Point hour hand at the sun. South is halfway between the hour hand and 12 o'clock position.

(c) Southern Hemisphere. Point the 12 o'clock position on your watch at the sun. North is halfway between the 12 o'clock position and the hour hand.

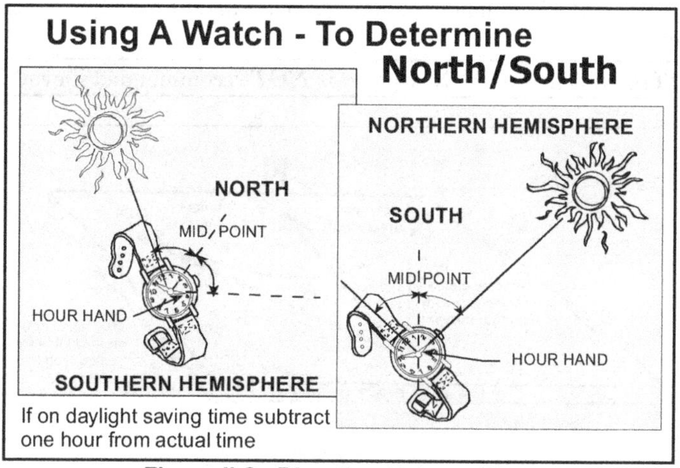


**Figure II-2. Direction Using a Watch**

(5) Using a pocket navigator **(Figure II-3)**—

(a) Gather the following necessary materials:

•Flat writing material (such as an MRE box).

•1-2 inch shadow tip device (a twig, nail, or match).

•Pen or pencil.

(b) Start construction at sunup; end construction at sundown. Do the following:

•Attach shadow tip device in center of paper.

•Secure navigator on flat surface (***DO NOT*** move during set up period).

•Mark tip of shadow every 30 minutes annotating the time.

•Connect marks to form an arc.

•Indicate north with a drawn arrow.

**Note**: The shortest line between base of shadow tip device and curved line is a north-south line.

(c) Do the following during travel:

•Hold navigator so the shadow aligns with mark of present time (drawn arrow now points to true north).

(d) Remember the navigator is current for approximately 1 week.

**CAUTION:** The Pocket Navigator is ***NOT*** recommended if evading.

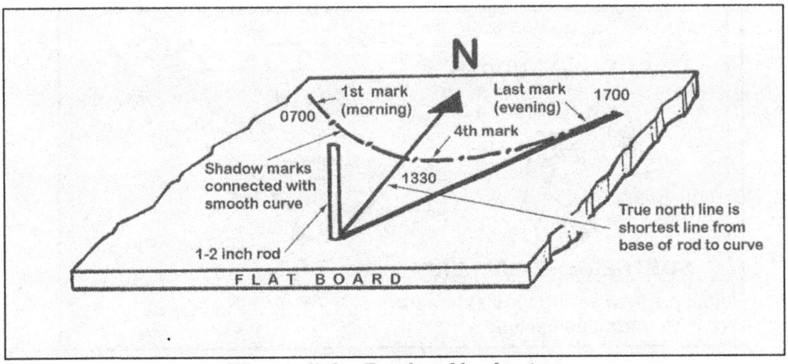


**Figure II-3. Pocket Navigator**

(6) Using the stars **(Figure II-4)** the—

(a) North Star is used to locate true north-south line.

(b) Southern Cross is used to locate true south-north line.

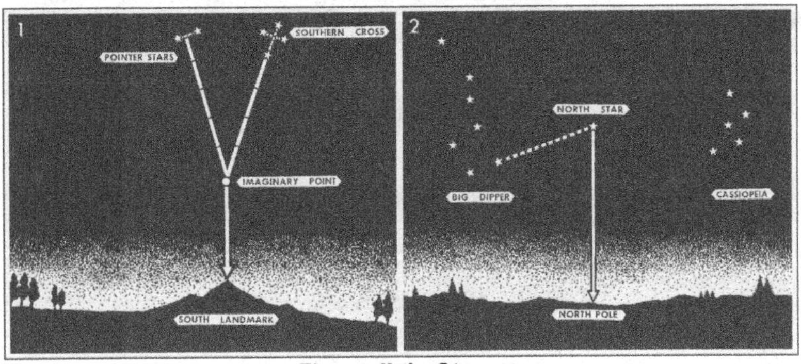


**Figure II-4. Stars**

c. Orient the map by—

(1) Using a true north-south line **(Figure II-5)**—

(a) Unfold map and place on a firm, flat, level nonmetallic surface.

(b) Align the compass on a true north-south line.

(c) Rotate map and compass until stationary index line aligns with the magnetic variation indicated in marginal information.

•Easterly (subtract variation from 360 degrees).

•Westerly (add variation to 360 degrees).

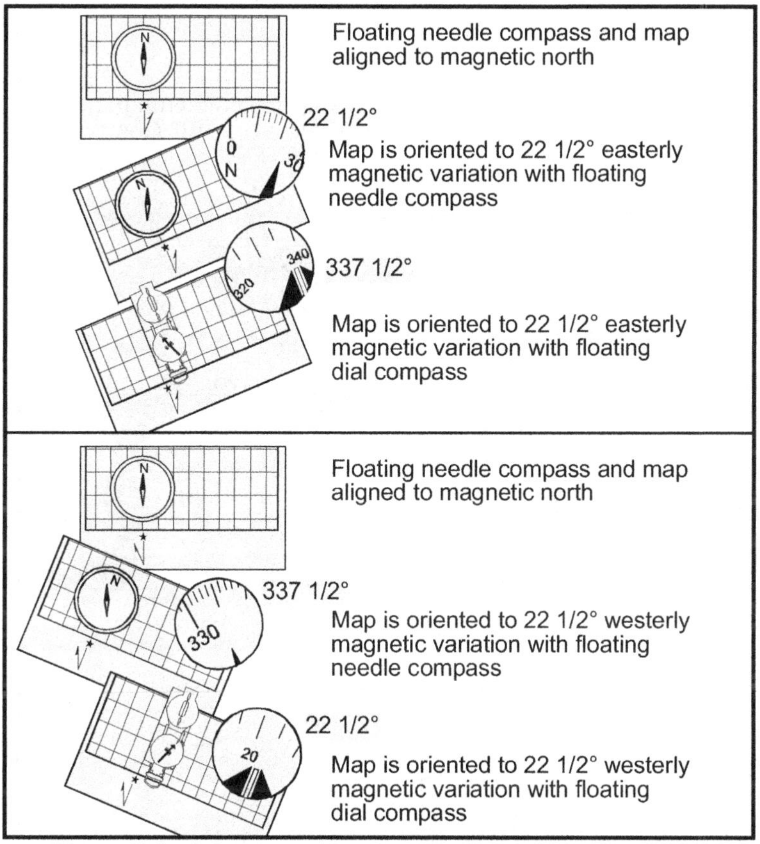


**Figure II-5. Orienting a Map Using a True North-South Line**

(2) Using a compass rose **(Figure II-6)**—

(a) Place edge of the lensatic compass on magnetic north line of the compass rose closest to your location.

(b) Rotate map and compass until compass reads 360 degrees.

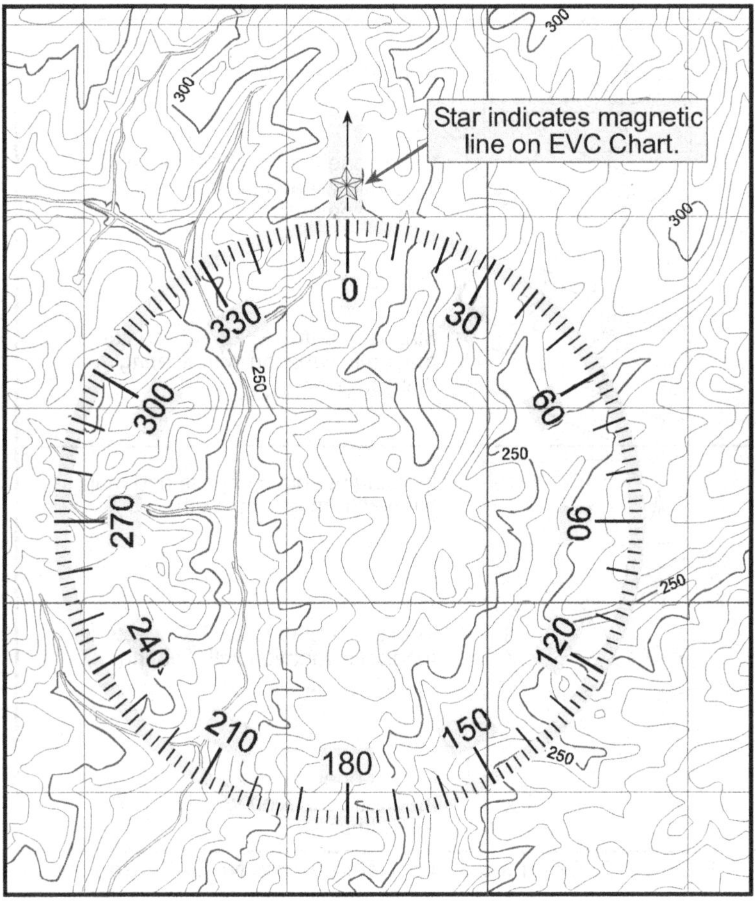


**Figure II-6. Map Orientation with Compass Rose**

(3) If there is **NO** compass, orient map using cardinal direction obtained by the stick and shadow method or the celestial aids (stars) method.

d. Determine specific location.

(1) Global Positioning System (GPS).

(a) ***DO NOT*** use GPS for primary navigation.

(b) Use GPS to confirm your position ***ONLY***.

(c) Select area providing maximum satellite reception.

(d) Conserve GPS battery life.

(2) Triangulation (resection) with a compass **(Figure II-7)**.

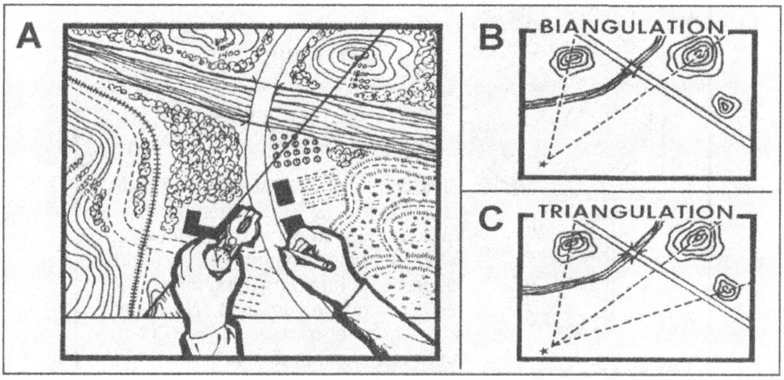


**Figure II-7. Triangulation**

(a) Try to use **3** or more azimuths.

(b) Positively identify a major land feature and determine a line of position (LOP).

(c) Check map orientation each time compass is used.

(d) Plot the LOP using a thin stick or blade of grass (combat) or pencil line (non-combat).

(e) Repeat steps **(b)** through **(d)** for other LOPs.

e. Use the compass for night navigation by—

(1) Setting up compass for night navigation **(Figure II-8)**.

(2) Aligning north-seeking arrow with luminous line and follow front of compass.

(3) Using point-to-point navigation.

f. Route selection techniques follow:

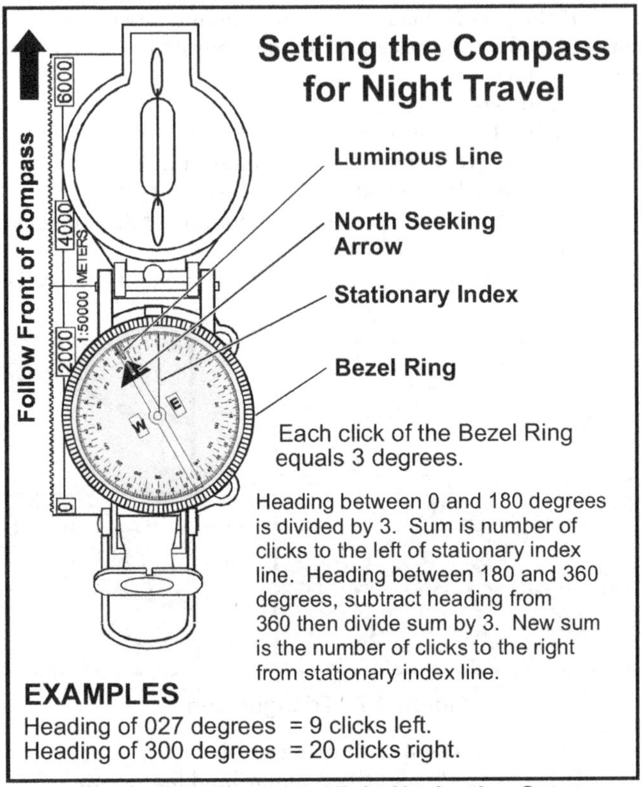


**Figure II-8. Compass Night Navigation Setup**

(1) Circumnavigation.

(a) Find a prominent landmark on the opposite side of the obstacle.

(b) Contour around obstacle to landmark.

(c) Resume your route of travel.

(2) Dogleg and 90 degree offset **(Figure II-9)**.

(3) Straight-line heading as follows:

(a) Maintain heading until reaching destination.

(b) Measure distance by counting the number of paces in a given course and convert to map units.

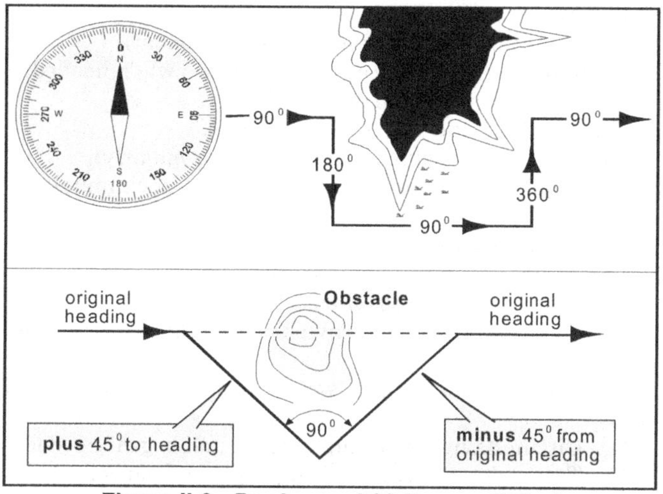


**Figure II-9. Dogleg and 90 Degree Offset**

•One pace is the distance covered each time the same foot touches the ground.

•Distances measured by paces are approximate (example in open terrain, 900 paces per kilometer [average], or example in rough terrain, 1200 paces per kilometer [average]).

(c) Use pace count in conjunction with terrain evaluation and heading to determine location. An individual's pace varies because of factors such as steep terrain, day/night travel, or injured/uninjured condition. Adjust estimation of distance traveled against these factors to get relative accuracy when using a pace count.

(4) Deliberate offset is—

(a) Used when finding a point on a linear feature (that is, road or river).

(b) Intentionally navigated to left or right of target so you know which way to turn at the linear feature.

(5) Point-to-point is same as straight line.

(a) Pick out landmarks on the heading and walk the trail of least resistance to a point.

(b) On reaching a point, establish another landmark and continue.

**3. Travel Considerations**

a. Pick the easiest and safest route (non-combat).

b. Maintain a realistic pace; take rest stops when needed.

c. Avoid overdressing and overheating.

d. Consider food and water requirements.

e. Take special care of feet (change socks regularly).

f. Pack equipment to prevent loss, damage, pack imbalance, and personal safety.

g. Go **around** obstacles, not over or through them.

h. Travel on trails whenever possible (non-combat).

i. Travel in forested areas if possible.

j. Avoid creek bottoms and ravines with ***NO*** escape in the event of heavy rains.

k. Consider the following for swamps, lakes, and unfordable rivers:

(1) Circumnavigate swamps, lakes, and bogs if needed.

(2) Travel downstream to find people and slower water.

(3) Travel upstream to find narrower and shallow water.

**4. River Travel**

River travel may be faster and save energy when hypothermia is not a factor. It may be a primary mode of travel and LOC in a tropical environment **(use with caution if evading)**.

a. Use flotation device (raft, log, bamboo, etc.).

b. Use a pole to move the raft in shallow water.

c. Use an oar in deep water.

d. Stay near inside edge of river bends (current speed is less).

e. Keep near shore.

f. Watch for the following ***DANGERS:***

(1) Snags.

(2) Sweepers (overhanging limbs and trees).

(3) Rapids (***DO NOT*** attempt to shoot the rapids).

(4) Waterfalls.

(5) Hazardous animals.

g. Consider using a flotation device when crossing rivers or large/deep streams.

## 5. Ice and Snow Travel

Travel should be limited to areas free of hazards.

a. ***DO NOT*** travel in—

(1) Blizzards.

(2) Bitterly cold winds.

(3) Poor visibility.

b. Obstacles to winter travel follow:

(1) Reduced daylight hours (***BE AWARE***).

(2) Deep soft snow (if movement is necessary, make snowshoes **[Figure II-10])**. Travel is easier in early morning or late afternoon near dusk when snow is frozen or crusted.

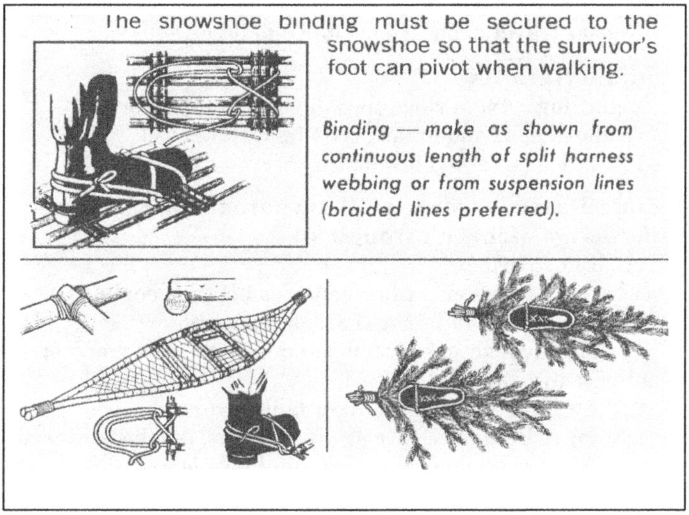


**Figure II-10. Improvised Snowshoes**

(3) Avalanche prone areas to avoid:

(a) Slopes 30-45 degrees or greater.

(b) Trees without uphill branches (identifies prior avalanches).

(c) Heavy snow loading on ridge tops.

(4) If caught in an avalanche, do the following:

(a) Backstroke to decrease burial depth.

(b) Move hand around face to create air pocket as moving snow slows.

(5) Frozen water crossings.

(a) Weak ice should be expected where—

•Rivers are straight.

•Objects protrude through ice.

•Snow banks extend over the ice.

•Rivers or streams come together.

•Water vapor rising indicates open or warm areas.

(b) Air pockets form when a frozen river loses volume.

(c) When crossing frozen water, distribute your weight by laying flat, belly crawling, or using snowshoes.

c. Glacier travel is hazardous and should be avoided.

**6. Mountain Hazards**

a. Lightning. Avoid ridge tops during thunderstorms.

b. Avalanche. Avoid areas prone to avalanches.

c. Flash floods. Avoid low areas.

**7. Summer Hazards (see page II-10; paragraph 3, *Travel Considerations*, items h through k.)**

(1) Dense brush.

(a) Travel on trails when possible (non-combat).

(b) Travel in forested areas if possible.

(c) Avoid creek bottoms and ravines with no escape in the event of heavy rains.

(2) Swamps, lakes, and unfordable rivers.

(a) Circumnavigate swamps, lakes, and bogs if needed.

(b) Travel downstream to find people and slower water.

(c) Travel upstream to find narrower and shallow water.

**8. Dry Climates**

a. ***DO NOT*** travel unless certain of reaching the destination using the water supply available.

b. Travel at dawn or dusk on hot days.

c. Follow the easiest trail possible (non-combat), avoiding—

(1) Deep sandy dune areas.

(2) Rough terrain.

d. In sand dune areas—

(1) Follow hard valley floor between dunes.

(2) Travel on the windward side of dune ridges.

e. If a sandstorm occurs—

(1) Mark your direction of travel.

(2) Sit or lie down in direction of travel.

(3) Try to get to the downwind side of natural shelter.

(4) Cover the mouth and nose with a piece of cloth.

(5) Protect the eyes.

(6) Remain stationary until the storm is over.

**9. Tropical Climates**

a. Travel only when it is light.

b. Avoid obstacles like thickets and swamps.

c. Part the vegetation to pass through. Avoid grabbing vegetation; it may have spines or thorns (**use gloves** if possible).

d. ***DO NOT*** climb over logs if you can go around them.

e. Find trails—

(1) Where **2** streams meet.

(2) Where a low pass goes over a range of hills.

f. While traveling trails—

(1) Watch for disturbed areas on game trails; they may indicate a pitfall or trap.

(2) Use a walking stick to probe for pitfalls or traps.

(3) ***DO NOT*** sleep on the trail.

(4) Exercise caution, the enemy uses the trails also.

**10. Open Seas**

a. Using currents—

(1) Deploy sea anchor **(Figure II-11)**. Sea anchor may be adjusted to make use of existing currents.

(2) Sit low in the raft.

(3) Deflate the raft slightly so it rides lower in the water.

b. Using winds—

(1) Pull in sea anchor.

(2) Inflate raft so it rides higher.

(3) Sit up in raft so body catches the wind.

(4) Construct a shade cover/sail **(Figure II-12)**. (Sail aids in making landfall.)

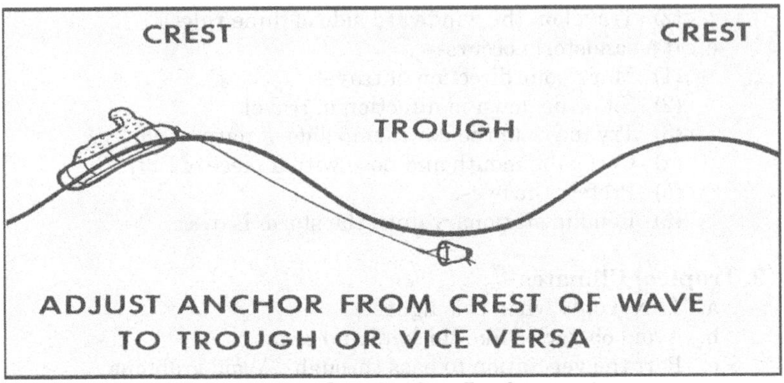

Figure II-11. Sea Anchor Deployment

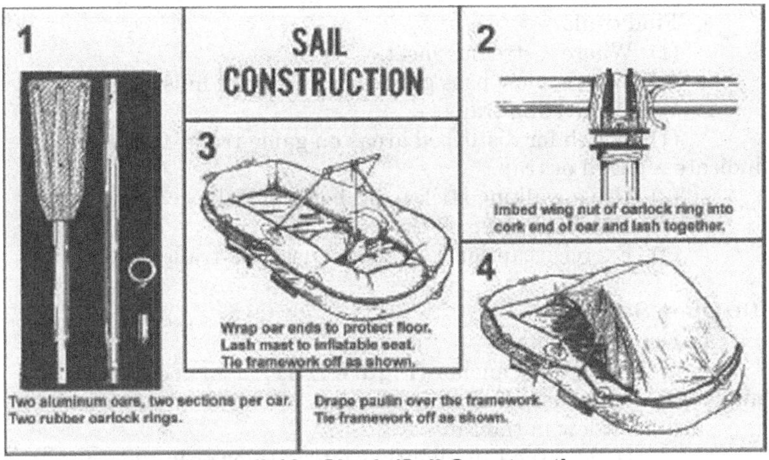

Figure II-12. Shade/Sail Construction

c. Making landfall. Indications of land are—

(1) Fixed cumulus clouds in a clear sky or in a cloudy sky where all other clouds are moving.

(2) Greenish tint in the sky **(in the tropics)**.

(3) Lighter colored reflection on clouds (open water causes dark gray reflections) (**in the arctic**).

(4) Lighter colored water (indicates shallow water).

(5) The odors and sounds.

(a) Odors from swamps and smoke.

(b) Roar of surf/bird cries coming from one direction.

(6) Directional flights of birds at dawn and at dusk.

d. Swimming ashore—

(1) Consider physical condition.

(2) Use a flotation aid.

(3) Secure all gear to body before reaching landfall.

(4) Remain in raft as long as possible.

(5) Use the sidestroke or breaststroke to conserve strength if thrown from raft.

(6) Wear footgear and at least **1** layer of clothing.

(7) Try to make landfall during the lull between the sets of waves (waves are generally in **sets** of **7**, from **smallest** to **largest**).

(8) In moderate surf.

(a) Swim forward on the back of a wave.

(b) Make a shallow dive just before the wave breaks to end the ride.

(9) In high surf.

(a) Swim shoreward in the trough between waves.

(b) When the seaward wave approaches, face it and submerge.

(c) After it passes, work shoreward in the next trough.

(10) If caught in the undertow of a large wave—

(a) Remain calm and swim to the surface.

(b) Lie as close to the surface as possible.

(c) Parallel shoreline and attempt landfall at a point further down shore.

(11) Select a landing point.

(a) Avoid places where waves explode upon rocks.

(b) Find a place where waves smoothly rush onto the rocks.

(12) After selecting a landing site—

(a) Face shoreward.

(b) Assume a sitting position with feet 2 or 3 feet lower than head to absorb the shock of hitting submerged objects.

e. Rafting ashore—

(1) Select landing point carefully.

(2) Use caution landing when the sun is low and straight in front of you causing poor visibility.

(3) Land on the lee (downwind) side of islands or point of land if possible.

(4) Head for gaps in the surf line.

(5) Penetrate surf by—

(a) Taking down most shade/sails.

(b) Using paddles to maintain control.

(c) Deploying a sea anchor for stability.

**CAUTION: *DO NOT*** deploy a sea anchor if traveling through coral.

f. Making sea ice landings on large stable ice flows. Icebergs, small flows, and disintegrating flows are dangerous **(ice can cut a raft)**.

(1) Use paddles to avoid sharp edges.

(2) Store raft away from the ice edge.

(3) Keep raft inflated and ready for use.

(4) Weight down/secure raft so it does not blow away.

## Chapter III

# RADIO COMMUNICATIONS AND SIGNALING

| |
|---|
| Inventory and review the operating instructions of all communications and signaling equipment. |

**1. Radio Communications (Voice and Data)**

a. Non-combat.

(1) Ensure locator beacon is operational.

(2) Follow standing plans for on/off operations to conserve battery use.

b. Combat.

(1) Turn off locator beacon.

(2) Keep it with you to supplement radio communications.

(3) Follow plans/orders for on/off operations.

c. Make initial contact as soon as possible or as directed in applicable plans/orders.

d. If no immediate contact, then as directed in applicable plans/orders.

e. Locate spare radio and batteries (keep warm and dry).

f. Transmissions.

(1) Use concealment sites (combat) that optimize line of site (LOS).

(2) Face recovery asset.

(3) Keep antenna perpendicular to intended receiver **(Figure III-1)**.

(4) ***DO NOT*** ground antenna (that is finger on antenna or attaching bolt, space blanket, vegetation, etc.).

(5) Keep transmissions short (3-5 seconds maximum). Use data burst if available.

(6) Move after each transmission (***ONLY*** in combat, if possible).

(7) If transmitting in the blind, ensure a clear LOS towards the equator.

(8) Use terrain masking to hinder enemy direction finding.

g. Listening (use reception times in applicable plans/orders or as directed by recovery forces).

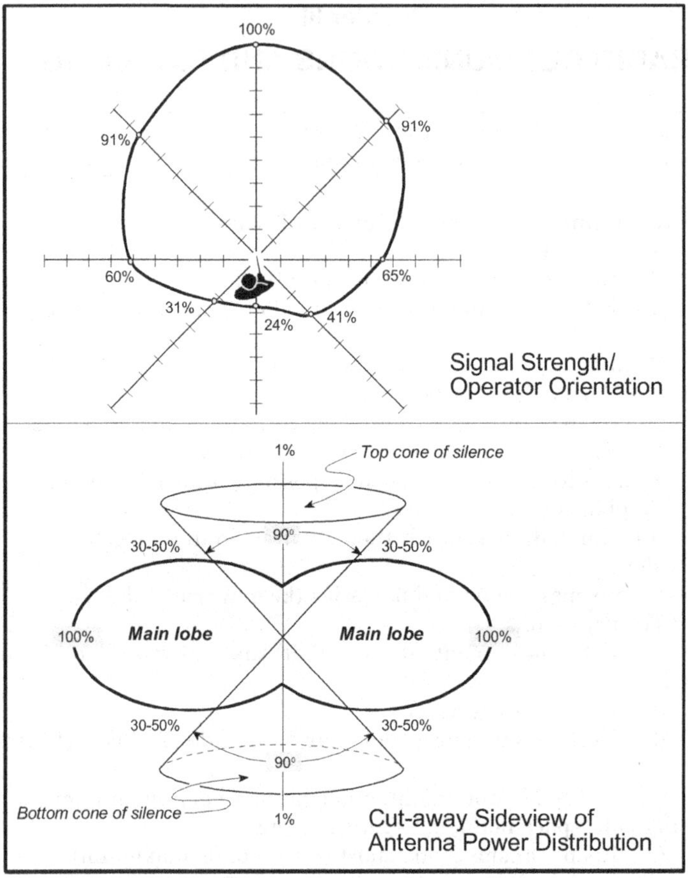


**Figure III-1. Radio Transmission Characteristics**

**2. Signaling**

a. Pyrotechnic signals.

(1) Prepare early (weather permitting).

(2) Use as directed in applicable plans/orders or as directed by recovery forces.

(3) Extend over raft's edge before activating.

b. Signal mirror **(Figure III-2)**.
(1) Use as directed by recovery forces.
(2) If no radio, use only with confirmed friendly forces.
(3) Cover when not in use.

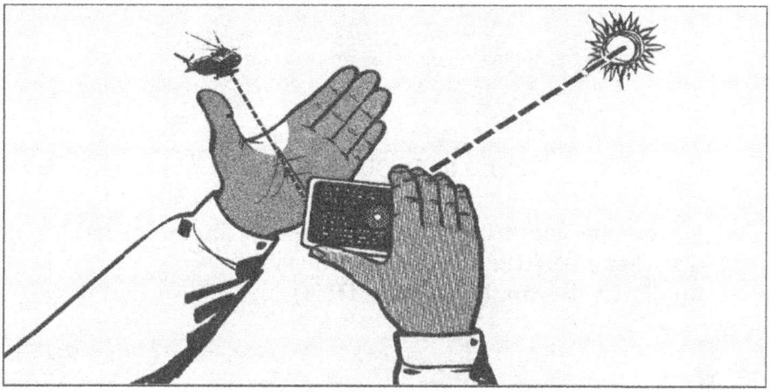

**Figure III-2. Sighting Techniques**

**Note:** Make a mirror from any shiny metal or glass.

c. Strobe/IR lights.
(1) Prepare early, consider filters and shields.
(2) Use as directed by recovery forces.
(3) Conserve battery life.

**Note:** Produces one residual flash when turned off.

d. Pattern signals (use as directed in applicable plans/orders).
(1) Materials:
(a) Manmade (space blanket, signal paulin, parachute).
(b) Natural use materials that contrast the color and/or texture of the signaling area (rocks, brush, branches, stomped grass).
(2) Location.
(a) Maximize visibility from above.
(b) Provide concealment from ground observation.
(3) Size (large as possible) and ratio **(Figure III-3)**.

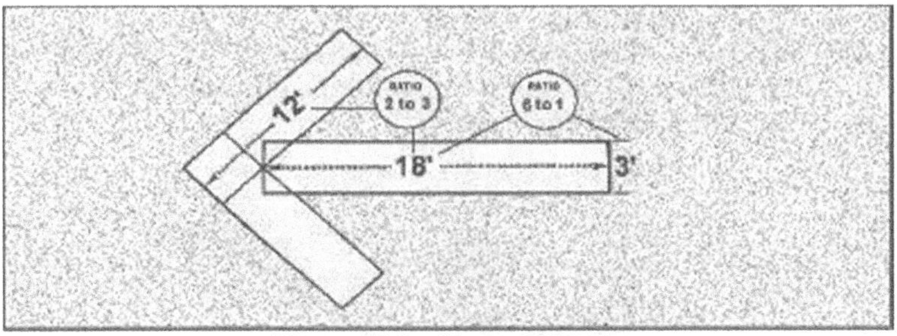


**Figure III-3. Size and Ratio**

(4) Shape (maintain straight lines and sharp corners).
(5) Contrast (use color and shadows).
(6) Pattern signals **(Figure III-4)**.

| NO. | MESSAGE | CODE SYMBOL |
|---|---|---|
| 1 | REQUIRE ASSISTANCE | V |
| 2 | REQUIRE MEDICAL ASSISTANCE | X |
| 3 | NO or NEGATIVE | N |
| 4 | YES or AFFIRMATIVE | Y |
| 5 | PROCEEDING IN THIS DIRECTION | ↑ |

**Figure III-4. Signal Key**

e. Sea dye marker.
(1) ***DO NOT*** waste in rough seas or fast moving water.
(2) Conserve unused dye by rewrapping.
(3) May be used to color snow.

f. Non-combat considerations:

(1) Use a fire at night.

(2) Use smoke for day (tires or petroleum products for dark smoke and green vegetation for light smoke). **(Figure III-5)**

(3) Use signal mirror to sweep horizon.

(4) Use audio signals (that is, voice, whistle, and weapons fire).

**Figure III-5. Smoke Generator**

## Chapter IV
# RECOVERY

**1. Responsibilities**

a. Establish radio contact with recovery forces (if possible).

b. Maintain communication with recovery forces until recovered.

c. Be prepared to authenticate as directed in applicable plans/orders.

d. Follow recovery force instructions, be prepared to report—

(1) Enemy activity in the recovery area.

(2) Recovery site characteristics (slope, obstacles, size, etc.).

(3) Number in party/medical situation.

(4) Signal devices available.

e. If no radio, a ground-to-air signal may be your only means to effect recovery.

**2. Site Selection**

a. Locate area for landing pick-up, if practical (approximately 150 feet diameter, free of obstructions, flat and level).

b. Assess evidence of human activity at/near the site (in combat).

c. Locate several concealment sites around area (in combat).

d. Plan several tactical entry and exit routes (in combat).

**3. Site Preparation**

a. Pack and secure all equipment.

b. Prepare signaling devices (use as directed or as briefed).

c. Mentally review recovery methods (aircraft, ground, boat, etc.).

**4. Recovery Procedures**

a. Assist recovery force in identifying your position.

b. Stay concealed until recovery is imminent (in combat).

c. For a landing/ground recovery—

(1) Assume a non-threatening posture.

(2) Secure weapons and avoid quick movement.

(3) ***DO NOT*** approach recovery vehicle until instructed.

(4) Beware of rotors/propellers when approaching recovery vehicle, especially on sloping or uneven terrain. Secure loose equipment that could be caught in rotors/propellers.

d. For **hoist** recovery devices **(Figures IV-1 and IV-2)**—

(1) Use eye protection, if available (glasses or helmet visor).

(2) Allow metal on device to contact the surface before touching to avoid injury from static discharge.

(3) Sit or kneel for stability while donning device.

(4) Put safety strap under armpits.

(5) Ensure cable is in front of you.

(6) Keep hands clear of all hardware and connectors.

(7) ***DO NOT*** become entangled in cable.

(8) Use a thumbs up, vigorous cable shake, or radio call to signal you are ready.

(9) Drag feet on the ground to decrease oscillation.

(10) ***DO NOT*** assist during hoist or when pulled into the rescue vehicle. Follow crewmember instructions.

e. For **nonhoist** recovery (rope or unfamiliar equipment)—

(1) Create a ***"fixed loop"*** big enough to place under armpits **(Figure IV-3)**.

(2) Follow the procedures in "**d**" above.

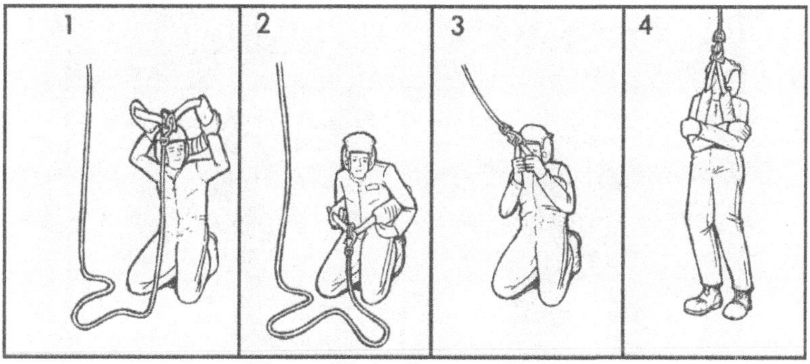


**Figure IV–1. Rescue Strap**

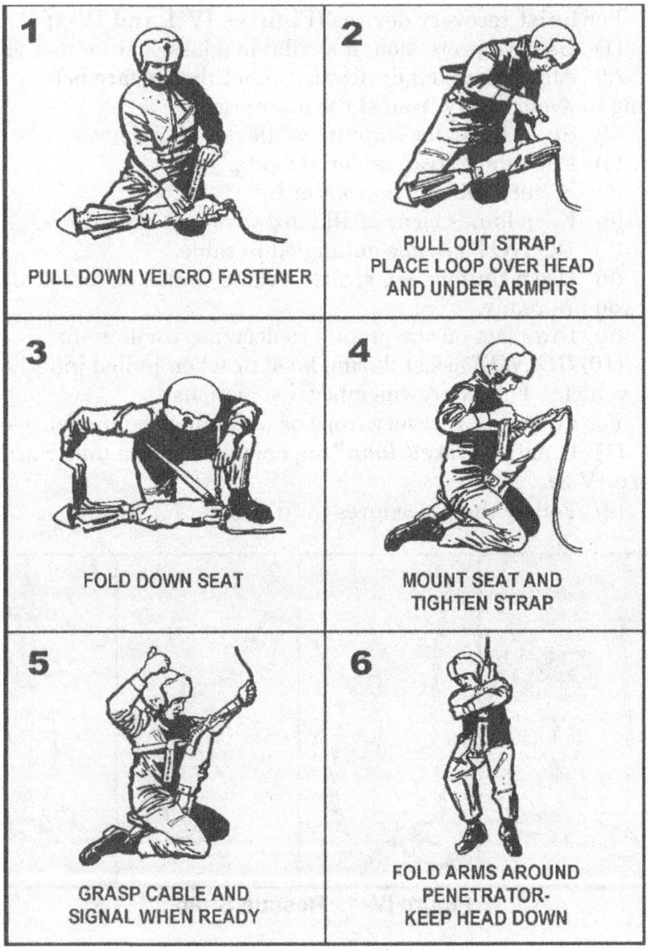


**Figure IV-2. Forest Penetrator**

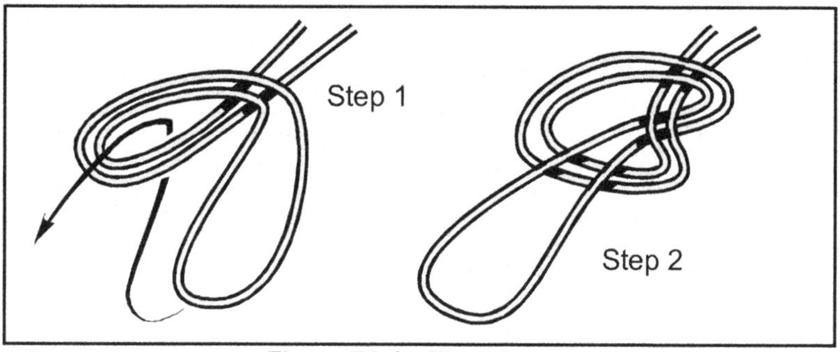

Figure IV–3. Fixed Loop

# Chapter V
# MEDICAL

**WARNING:** These emergency medical procedures are for survival situations. Obtain professional medical treatment as soon as possible.

## 1. Immediate First Aid Actions

**Remember the *ABCs* of Emergency Care:**
Airway Breathing Circulation

a. Determine responsiveness as follows:
  (1) If unconscious, arouse by shaking gently and shouting.
  (2) If no response—
    (a) Keep head and neck aligned with body.
    (b) Roll victims onto their backs.
    (c) Open the airway by lifting the chin **(Figure V-1)**.
    (d) Look, listen, and feel for air exchange.

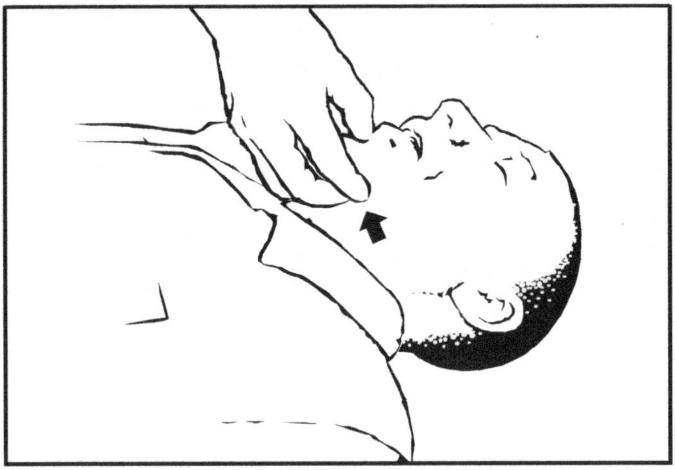

**Figure V-1. Chin Lift**

(3) If victim is not breathing—
(a) Check for a clear airway; remove any blockage.
(b) Cover victim's mouth with your own.
(c) Pinch victim's nostrils closed.
(d) Fill victim's lungs with **2** slow breaths.
(e) If breaths are blocked, reposition airway; try again.
(f) If breaths still blocked, give **5** abdominal thrusts:
•Straddle the victim.
•Place a fist between breastbone and belly button.
•Thrust upward to expel air from stomach.
(g) Sweep with finger to clear mouth.
(h) Try **2** slow breaths again.
(i) If the airway is still blocked, continue **(c)** through **(f)** until successful or exhausted.
(j) With open airway, start mouth to mouth breathing:
•Give **1** breath every 5 seconds.
•Check for chest rise each time.

(4) If victim is unconscious, but breathing—
(a) Keep head and neck aligned with body.
(b) Roll victim on side (drains the mouth and prevents the tongue from blocking airway).

(5) If breathing difficulty is caused by chest trauma, refer to **page V-7, paragraph 1d, *Treat Chest Injuries*.**

**CAUTION:** ***DO NOT*** remove an impaled object unless it interferes with the airway. You may cause more tissue damage and increase bleeding. For travel, you may shorten and secure the object.

b. Control bleeding as follows:
(1) Apply a pressure dressing **(Figure V-2)**.
(2) If ***STILL*** bleeding—
(a) Use direct pressure over the wound.
(b) Elevate the wounded area above the heart.

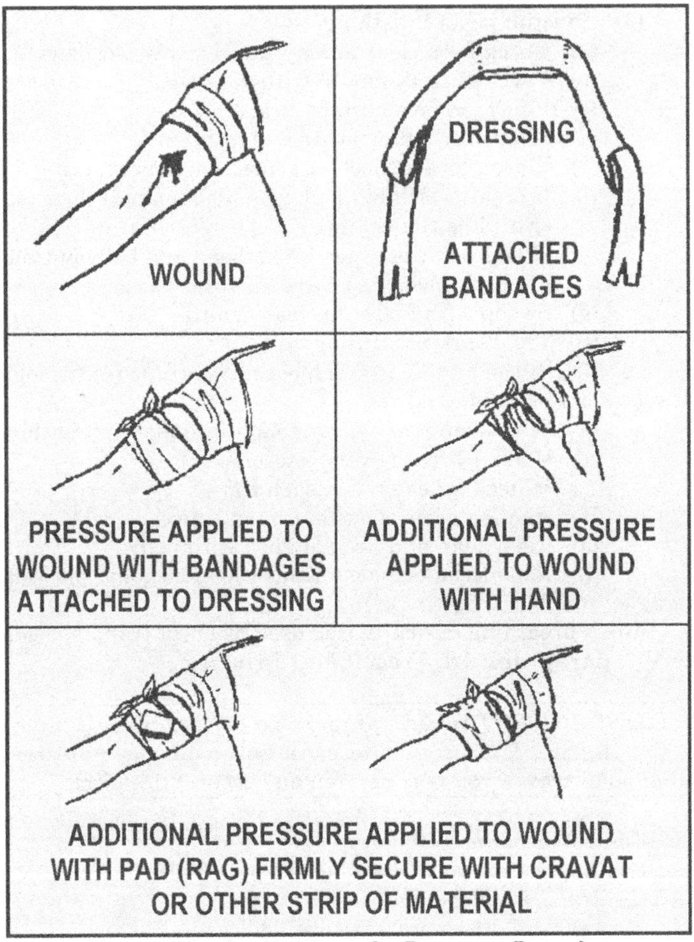


**Figure V-2. Application of a Pressure Dressing**

(3) If ***STILL*** bleeding—

(a) Use a pressure point between the injury and the heart **(Figure V-3)**.

(b) Maintain pressure for 6 to 10 minutes before checking to see if bleeding has stopped.

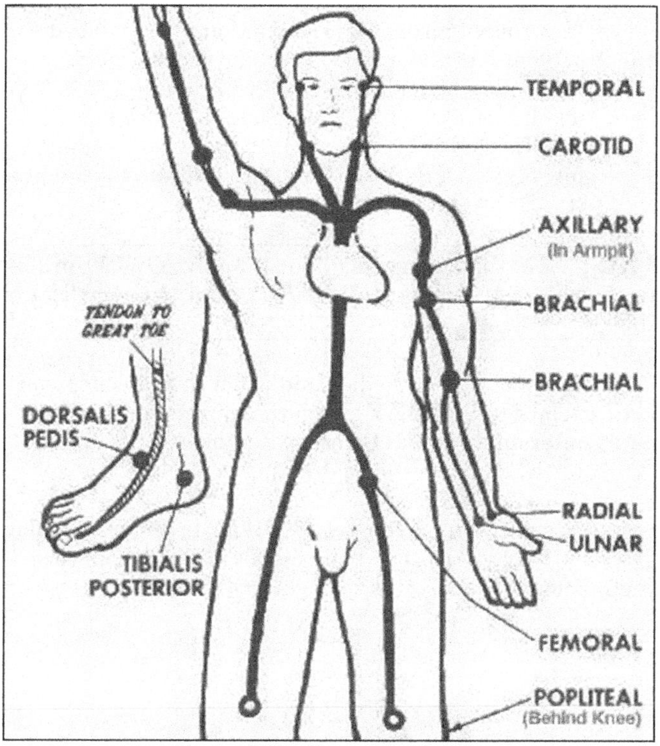


**Figure V-3. Pressure Points**

(4) If a limb wound is ***STILL*** bleeding—

**CAUTION:** Use of a tourniquet is a ***LAST RESORT*** measure. Use ***ONLY*** when severe, uncontrolled bleeding will cause loss of life. Recognize that long-term use of a tourniquet may cause loss of limb.

(a) Apply tourniquet (TK) band just above bleeding site on limb. A band at least 3 inches (7.5 cm) or wider is best.

(b) Follow steps illustrated in **Figure V-4**.

(c) Use a stick at least 6 inches (15 cm) long.

(d) Tighten only enough to stop arterial bleeding.

(e) Mark a ***TK*** on the forehead with the time applied.

(f) ***DO NOT*** cover the tourniquet.

**CAUTION:** The following directions apply ***ONLY*** in survival situations where rescue is ***UNLIKELY*** and ***NO*** medical aid is available.

(g) If rescue or medical aid is not available for over 2 hours, an attempt to ***SLOWLY*** loosen the tourniquet may be made 20 minutes after application. Before loosening—

- Ensure pressure dressing is in place.
- Ensure bleeding has stopped
- Loosen tourniquet ***SLOWLY*** to restore circulation.
- Leave loosened tourniquet in position in case bleeding resumes.

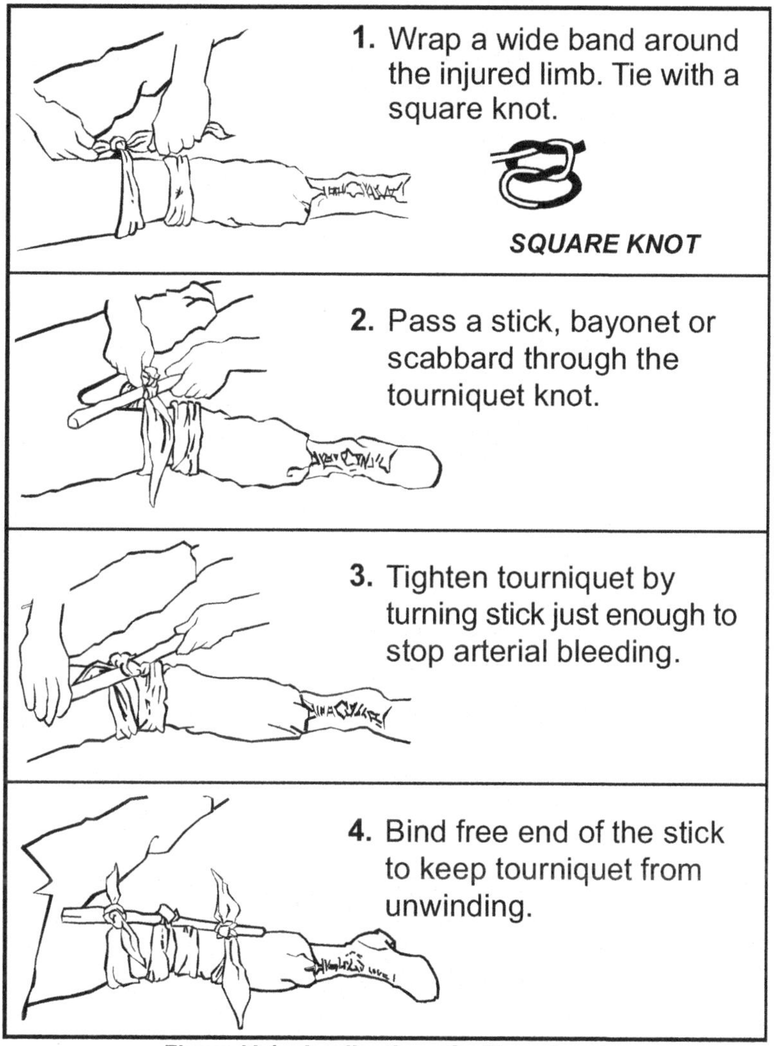


**Figure V-4. Application of a Tourniquet**

c. Treat shock. (Shock is difficult to identify or treat under field conditions. It may be present with or without visible injury.)

(1) Identify by one or more of the following:

(a) Pale, cool, and sweaty skin.

(b) Fast breathing and a weak, fast pulse.

(c) Anxiety or mental confusion.

(d) Decreased urine output.

(2) Maintain circulation.

(3) Treat underlying injury.

(4) Maintain normal body temperature.

(a) Remove wet clothing.

(b) Give warm fluids.

•***DO NOT*** give fluids to an unconscious victim.

•***DO NOT*** give fluids if they cause victim to gag.

(c) Insulate from ground.

(d) Shelter from the elements.

(5) Place conscious victim on back.

(6) Place very weak or unconscious victim on side, this will—

(a) Allow mouth to drain.

(b) Prevent tongue from blocking airway.

d. Treat chest injuries.

(1) Sucking chest wound. This occurs when chest wall is penetrated; may cause victim to gasp for breath; may cause sucking sound; may create bloody froth as air escapes the chest.

(a) ***Immediately*** seal wound with hand or airtight material.

(b) Tape airtight material over wound on ***3 sides only*** **(Figure V-5)** to allow air to escape from the wound but not to enter.

(c) Monitor breathing and check dressing.

(d) Lift untapped side of dressing as victim ***<u>exhales</u>*** to allow trapped air to escape, as necessary.

(2) Flail chest. Results from blunt trauma when ***3*** or ***more*** ribs are broken in ***2*** or more places. The flail segment is the broken area that moves in a direction opposite to the rest of chest during breathing.

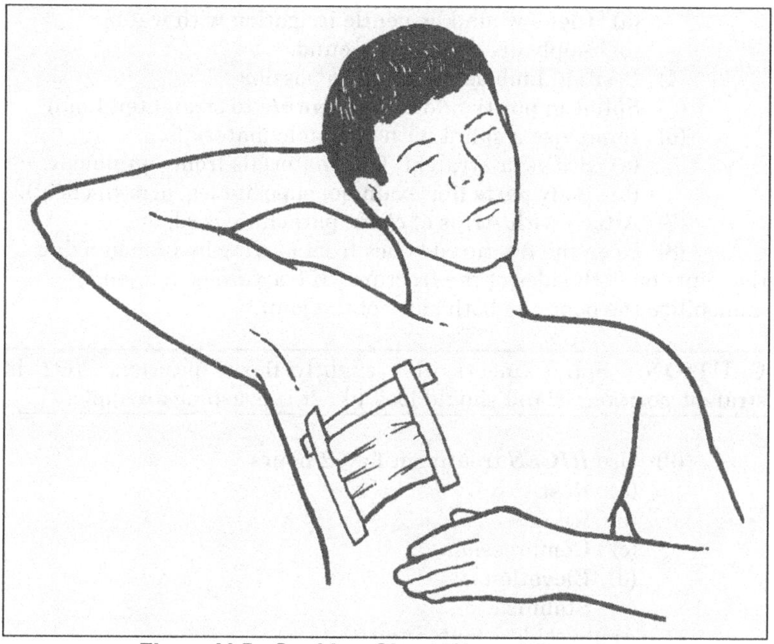

**Figure V-5. Sucking Chest Wound Dressing**

(a) Stabilize the flail segment as follows:

•Place rolled-up clothing or bulky pad over site.

•Tape pad to site

•***DO NOT*** wrap tape around chest.

(b) Have victim keep segment still with hand pressure.

(c) Roll victim onto side of flail segment injury (as other injuries allow).

(3) Fractured ribs.

(a) Encourage deep breathing (painful, but necessary to prevent the possible development of pneumonia).

(b) ***DO NOT*** constrict breathing by taping ribs.

e. Treat fractures, sprains, and dislocations.

(1) Control bleeding.

(2) Remove watches, jewelry, and constrictive clothing.

(3) If fracture penetrates the skin—

(a) Clean wound by gentle irrigation with water.
(b) Apply dressing over wound.
(4) Position limb as normally as possible.
(5) Splint in position found (if ***unable*** to straighten limb).
(6) Improvise a splint with available materials:
(a) Sticks or straight, stiff materials from equipment.
(b) Body parts (for example, opposite leg, arm-to-chest).
(7) Attach with strips of cloth, parachute cord, etc.
(8) Keep the fractured bones from moving by immobilizing the joints on both sides of the fracture. If fracture is in a joint, immobilize the bones on both sides of the joint.

**CAUTION:** Splint fingers in a slightly flexed position, ***NOT*** in straight position. Hand should look like it is grasping an apple.

(9) Use ***RICES*** treatment for 72 hours.
(a) **R**est.
(b) **I**ce.
(c) **C**ompression.
(d) **E**levation.
(e) **S**tabilization.
(10) Apply cold to acute injuries.
(11) Use 15 to 20 minute periods of cold application.
(a) ***DO NOT*** use continuous cold therapy.
(b) Repeat 3 to 4 times per day.
(c) Avoid cooling that can cause frostbite or hypothermia.
(12) Wrap with a compression bandage after cold therapy.
(13) Elevate injured area above heart level to reduce swelling.
(14) Check periodically for a pulse beyond the injury site.
(15) Loosen bandage or reapply splint if no pulse is felt or if swelling occurs because bandage is too tight.

**2. Common Injuries and Illnesses**
a. Burns.
(1) Cool the burned area with water.
(a) Use immersion or cool compresses.
(b) Avoid aggressive cooling with ice or frigid water.
(2) Remove watches, jewelry, constrictive clothing.

(3) ***DO NOT*** remove embedded, charred material that will cause burned areas to bleed.

(4) Cover with sterile dressings.

(5) ***DO NOT*** use lotion or grease.

(6) Avoid moving or rubbing the burned part.

(7) Drink ***extra*** water to compensate for increased fluid loss from burns. (Add ***1/4 teaspoon*** of ***salt*** [if available] to ***each quart*** of ***water***.)

(8) Change dressings when soaked or dirty.

b. Eye injuries.

(1) Sun/snow blindness (gritty, burning sensation, and possible reduction in vision caused by sun exposure).

(a) Prevent with improvised goggles. **(See Chapter VI, page VI-3, Figure VI-2.)**

(b) Treat by patching affected eye(s).

- Check after 12 hours.
- Replace patch for another 12 hours if not healed.

(c) Use cool compresses to reduce pain.

(2) Foreign body in eye.

(a) Irrigate with clean water from the ***inside*** to the ***outside*** corner of the eye.

(b) If foreign body is not removed by irrigation, improvise a small swab. Moisten and wipe gently over the affected area.

(c) If foreign body is ***STILL*** not removed, patch eye for 24 hours and then reattempt removal using steps **(a)** and **(b)**.

c. Heat injury.

(1) Heat cramps (cramps in legs or abdomen).

(a) Rest.

(b) Drink water. Add ***1/4 teaspoon*** of salt ***per quart***.

(2) Heat exhaustion (pale, sweating, moist, cool skin).

(a) Rest in shade.

(b) Drink water.

(c) Protect from further heat exposure.

(3) Heat stroke (victim disoriented or unconscious, skin is hot and flushed [sweating **may** or **may not** occur], fast pulse).

**CAUTION:** Handle heat stroke victim gently. Shock, seizures, and cardiac arrest can occur.

(a) Cool as rapidly as possible (saturate clothing with water and fan the victim). Remember to cool the groin and armpit areas. (Avoid overcooling.)

(b) Maintain airway, breathing, and circulation.

d. Cold injuries:

(1) Frostnip and frostbite—

(a) Are progressive injuries.

•Ears, nose, fingers, and toes are affected first.

•Areas will feel cold and may tingle leading to—

••Numbness that progresses to—

•••Waxy appearance with stiff skin that cannot glide freely over a joint.

(b) Frostnipped areas rewarm with body heat. If body heat ***WILL NOT*** rewarm area in 15 to 20 minutes, then frostbite is present.

(c) Frostbitten areas are deeply frozen and require medical treatment.

**CAUTION:** In frostbite, repeated freezing and thawing causes severe pain and increases damage to the tissue. ***DO NOT*** rub frozen tissue. ***DO NOT*** thaw frozen tissue.

(2) Hypothermia—

(a) Is a progressive injury.

•Intense shivering with impaired ability to perform complex tasks leads to—

••Violent shivering, difficulty speaking, sluggish thinking go to—

•••Muscular rigidity with blue, puffy skin; jerky movements go to—

••••Coma, respiratory and cardiac failure.

(b) Protect victim from the environment as follows:

•Remove wet clothing.

•Put on dry clothing (if available).

•Prevent further heat loss.

••Cover top of head.

••Insulate from above and below.

•Warm with blankets, sleeping bags, or shelter.

•Warm central areas before extremities.
••Place heat packs in groin, armpits, and around neck.
••Avoid causing burns to skin.

**CAUTION**: Handle hypothermia victim gently. Avoid overly rapid rewarming which may cause cardiac arrest. Rewarming of victim with skin-to-skin contact by volunteer(s) inside of a sleeping bag is a survival technique but can cause internal temperatures of all to drop.

e. Skin tissue damage.
(1) Immersion injuries. Skin becomes wrinkled as in ***dishpan hands***.
(a) Avoid walking on affected feet.
(b) Pat dry; ***DO NOT*** rub. Skin tissue will be sensitive.
(c) Dry socks and shoes. Keep feet protected.
(d) Loosen boots, cuffs, etc., to improve circulation.
(e) Keep area dry, warm, and open to air.
(f) ***DO NOT*** apply creams or ointments.
(2) Saltwater sores.
(a) Change body positions frequently.
(b) Keep sores dry.
(c) Use antiseptic (if available).
(d) ***DO NOT*** open or squeeze sores.
f. Snakebite.

**CAUTION:** This snakebite treatment recommendation is for situations where medical aid and specialized equipment are not available.

(1) Nonpoisonous. Clean and bandage wound.
(2) Poisonous.
(a) Remove constricting items.
(b) Minimize activity.
(c) ***DO NOT*** cut the bite site; ***DO NOT*** use your mouth to create suction.
(d) Clean bite with soap and water; cover with a dressing.

(e) Overwrap the bite site with a tight (elastic) bandage **(Figure V-6)**. The intent is to slow capillary and venous blood flow but not arterial flow. Check for pulse below the overwrap.

(f) Splint bitten extremity to prevent motion.

(g) Treat for shock **(page V-7, paragraph 1c)**.

(h) Position extremity below level of heart.

(i) Construct shelter if necessary (let the victim rest).

(j) For conscious victims, force fluids.

g. Marine life.

(1) Stings.

(a) Flush wound with salt water (fresh water stimulates toxin release).

(b) Remove jewelry and watches.

(c) Remove tentacles and gently scrape or shave skin.

(d) Apply a steroid cream (if available).

(e) ***DO NOT*** rub area with sand.

(f) Treat for shock; artificial respiration may be required **(page V-1, paragraph 1a)**.

(g) ***DO NOT*** use urine to flush or treat wounds.

(2) Punctures.

(a) Immerse affected part in hot water or apply hot compresses for 30-60 minutes (as hot as victim can tolerate).

(b) Cover with clean dressing.

(c) Treat for shock as needed.

h. Skin irritants (includes poison oak and poison ivy).

(1) Wash with large amounts of water. Use soap (if available).

(2) Keep covered to prevent scratching.

i. Infection.

(1) Keep wound clean.

(2) Use iodine tablet solution or diluted betadine to prevent or treat infection.

(3) Change bandages as needed.

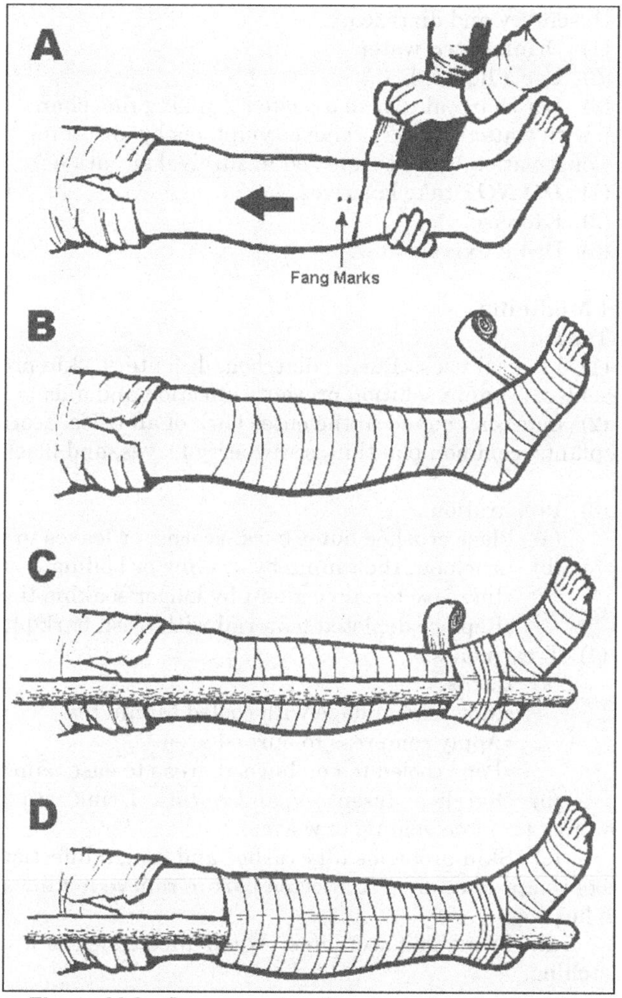

Figure V-6. Compression Bandage for Snake Bite

j. Dysentery and diarrhea.

(1) Drink ***extra*** water.

(2) Use a liquid diet.

(3) Eat charcoal. Make a paste by mixing fine charcoal particles with water. (It may relieve symptoms by absorbing toxins.)

k. Constipation (can be expected in survival situations).

(1) ***DO NOT*** take laxatives.

(2) Exercise.

(3) Drink **extra** water.

**3. Plant Medicine**

a. Tannin.

(1) Medical uses. Burns, diarrhea, dysentery, skin problems, and parasites. Tannin solution prevents infection and aids healing.

(2) Sources. Found in the outer bark of all trees, acorns, banana plants, common plantain, strawberry leaves, and blackberry stems.

(3) Preparation.

(a) Place crushed outer bark, acorns, or leaves in water.

(b) Leach out the tannin by soaking or boiling.

- Increase tannin content by longer soaking time.
- Replace depleted material with fresh bark/plants.

(4) Treatments.

(a) Burns.

- Moisten bandage with cooled tannin tea.
- Apply compress to burned area.
- Pour cooled tea on burned areas to ease pain.

(b) Diarrhea, dysentery, and worms. Drink strong tea solution (may promote voiding of worms).

(c) Skin problems (dry rashes and fungal infections). Apply cool compresses or soak affected part to relieve itching and promote healing.

(d) Lice and insect bites. Wash affected areas with tea to ease itching.

b. Salicin/salicylic acid.

(1) Medical uses. Aches, colds, fever, inflammation, pain, sprains, and sore throat (aspirin-like qualities).

(2) Sources. Willow and aspen trees **(Figure V-7)**.

(3) Preparation.

(a) Gather twigs, buds, or cambium layer (soft, moist layer between the outer bark and the wood) of willow or aspen.

(b) Prepare tea as described in paragraph **3a(3)**.

(c) Make poultice.

•Crush the plant or stems.

•Make a pulpy mass.

(4) Treatments.

(a) Chew on twigs, buds, or cambium for symptom relief.

(b) Drink tea for colds and sore throat.

(c) Use warm, moist poultice for aches and sprains.

•Apply pulpy mass over injury.

•Hold in place with a dressing.

c. Common plantain.

(1) Medical uses. Itching, wounds, abrasions, stings, diarrhea, and dysentery.

(2) Source. There are over 200 plantain species with similar medicinal properties. The common plantain is shown in **Figure V-7**.

(3) Preparation.

(a) Brew tea from seeds.

(b) Brew tea from leaves.

(c) Make poultice of leaves.

(4) Treatments.

(a) Drink tea made from seeds for diarrhea or dysentery.

(b) Drink tea made from leaves for vitamin and minerals.

(c) Use poultice to treat cuts, sores, burns, and stings.

d. Papain.

(1) Medical uses. Digestive aid, meat tenderizer, and a food source.

(2) Source. Fruit of the papaya tree **(Figure V-7)**.

(3) Preparation.

(a) Make cuts in **unripe** fruit.

(b) Gather milky white sap for its papain content.

(c) Avoid getting sap in eyes or wounds.

(4) Treatments.

(a) Use sap to tenderize tough meat.

(b) Eat **ripe** fruit for food, vitamins, and minerals.

e. Common Cattail.

(1) Medical uses. Wounds, sores, boils, inflammations, burns, and an excellent food source.

(2) Source. Cattail plant found in marshes **(Figure V-7)**.

(3) Preparation.

(a) Pound roots into a pulpy mass for a poultice.

(b) Cook and eat green bloom spikes.

(c) Collect yellow pollen for flour substitute.

(d) Peel and eat tender shoots (raw or cooked).

(4) Treatments.

(a) Apply poultice to affected area.

(b) Use plant for food, vitamins, and minerals.

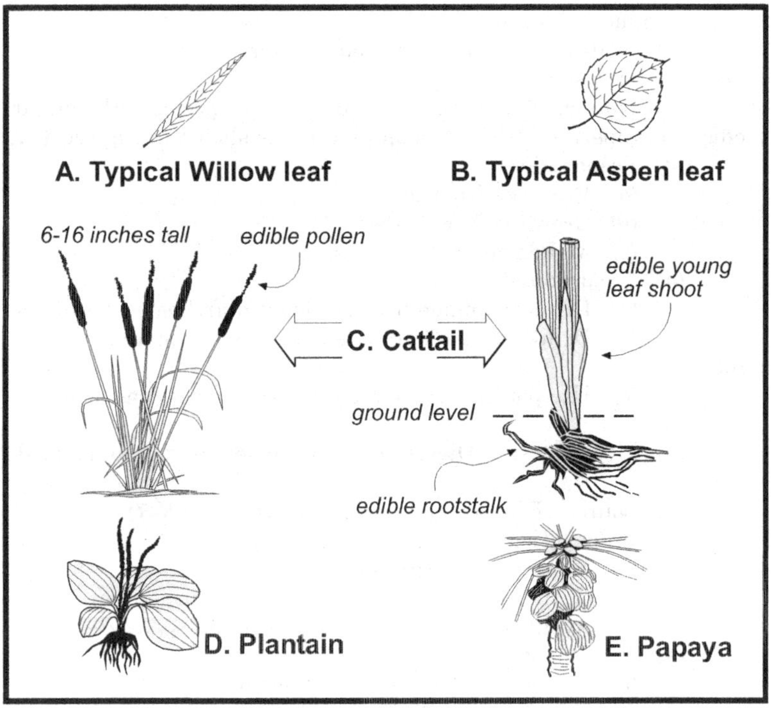


**Figure V-7. Useful Plants**

**4. Health and Hygiene**

a. Stay clean (daily regimen).

(1) Minimize infection by washing. (Use white ashes, sand, or loamy soil as soap substitutes.)

(2) Comb and clean debris from hair.

(3) Cleanse mouth and brush teeth.

(a) Use hardwood twig as toothbrush (fray it by chewing on one end then use as brush).

(b) Use single strand of an inner core string from parachute cord for dental floss.

(c) Use clean finger to stimulate gum tissues by rubbing.

(d) Gargle with salt water to help prevent sore throat and aid in cleaning teeth and gums.

(4) Clean and protect feet.

(a) Change and wash socks

(b) Wash, dry, and massage.

(c) Check frequently for blisters and red areas.

(d) Use adhesive tape/mole skin to prevent damage.

b. Exercise daily.

c. Prevent and control parasites.

(1) Check body for lice, fleas, ticks, etc.

(a) Check body regularly.

(b) Pick off insects and eggs (***DO NOT*** crush).

(2) Wash clothing and use repellents.

(3) Use smoke to fumigate clothing and equipment.

**5. Rules for Avoiding Illness**

a. Purify all water obtained from natural sources by using iodine tablets, bleach, or boiling for 5 minutes.

b. Locate latrines 200 feet from water and away from shelter.

c. Wash hands before preparing food or water.

d. Clean all eating utensils after each meal.

e. Prevent insect bites by using repellent, netting, and clothing.

f. Dry wet clothing as soon as possible.

g. Eat varied diet.

h. Try to get 7-8 hours sleep per day.

## Chapter VI
# PERSONAL PROTECTION

**1. Priorities**

a. Evaluate available resources and situation, then accomplish individual tasks accordingly.

b. First 24 hours in order of situational needs—

(1) Construct survival shelter according to selection criteria.

(2) Procure water.

(3) Establish multiple survival signals.

(4) Build Fire.

c. Second 24 hours in order of situational needs—

(1) Construct necessary tools and weapons.

(2) Procure food.

**2. Care and Use of Clothing**

a. Never discard clothing.

b. Wear loose and layered clothing.

(1) Tight clothing restricts blood flow regulating body temperature.

(2) Layers create more dead air space.

c. Keep entire body covered to prevent sunburn and dehydration in hot climates. When fully clothed, the majority of body heat escapes through the head and neck areas.

d. Avoid overheating.

(1) Remove layers of clothing before strenuous activities.

(2) Use a hat to regulate body heat.

(3) Wear a hat when in direct sunlight (in hot environment).

e. Dampen clothing when on the ocean in hot weather.

(1) Use salt water, ***NOT*** drinking water.

(2) Dry clothing before dark to prevent hypothermia.

f. Keep clothing dry to maintain its insulation qualities (dry damp clothing in the sun or by a fire).

g. If you fall into the water in the winter—

(1) Build fire.

(2) Remove wet clothing and rewarm by fire.

(3) Finish drying clothing by fire.

h. If no fire is available—
   (1) Remove clothing and get into sleeping bag (if available).
   (2) Allow wet clothes to freeze.
   (3) Break ice out of clothing.

i. Keep clothing clean (dirt reduces its insulation qualities). Examine clothing frequently for damage.
   (1) ***DO NOT*** sit or lie directly on the ground.
   (2) Wash clothing whenever possible.
   (3) Repair when necessary by using—
      (a) Needle and thread.
      (b) Safety pins.
      (c) Tape.

j. Improvised foot protection **(Figure VI-1)**.
   (1) Cut **2** to **4** layers of cloth into a 30-inch square.
   (2) Fold into a triangle.
   (3) Center foot on triangle with toes toward corner.

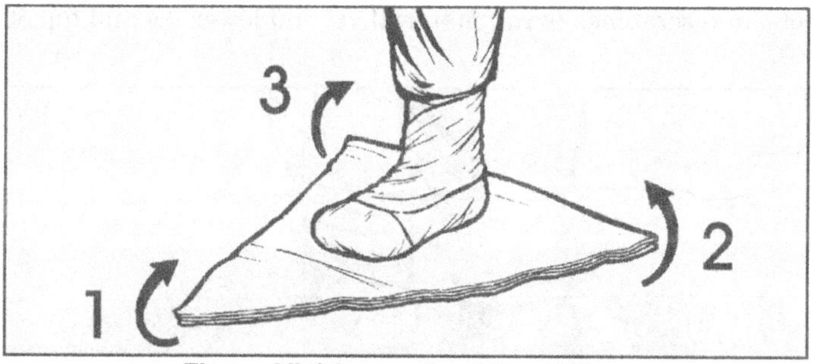


**Figure VI-1 Improvised Foot Wear**

   (4) Fold front over the toes.
   (5) Fold side corners, one at a time, over the instep.
   (6) Secure by rope, vines, tape, etc., or tuck into other layers of material.

**3. Other Protective Equipment**

a. Sleeping bag.
   (1) Fluff before use, ***especially*** at foot of bag.
   (2) Air and dry daily to remove body moisture.

(3) Improvise with available material, dry grass, leaves, dry moss, etc.

b. Sun and snow goggles **(Figure VI-2)**.

(1) Wear in bright sun or snow conditions.

(2) Improvise by cutting small horizontal slits in webbing, bark, or similar materials.

**Figure VI-2. Sun and Snow Goggles**

c. Gaiters **(Figure VI-3)**. Used to protect from sand, snow, insects, and scratches (wrap material around lower leg and top of boots).

**Figure VI-3. Gaiters**

**4. Shelters**

Evasion considerations apply.

a. Site selection.

(1) Near signal and recovery site.

(2) Available food and water.

(3) Avoid natural hazards:

(a) Dead standing trees.

(b) Drainage and dry river beds except in combat areas.

(c) Avalanche areas.

(4) Location large and level enough to lie down in.

b. Types.

(1) Immediate shelters. Find shelter needing minimal improvements **(Figure VI-4)**.

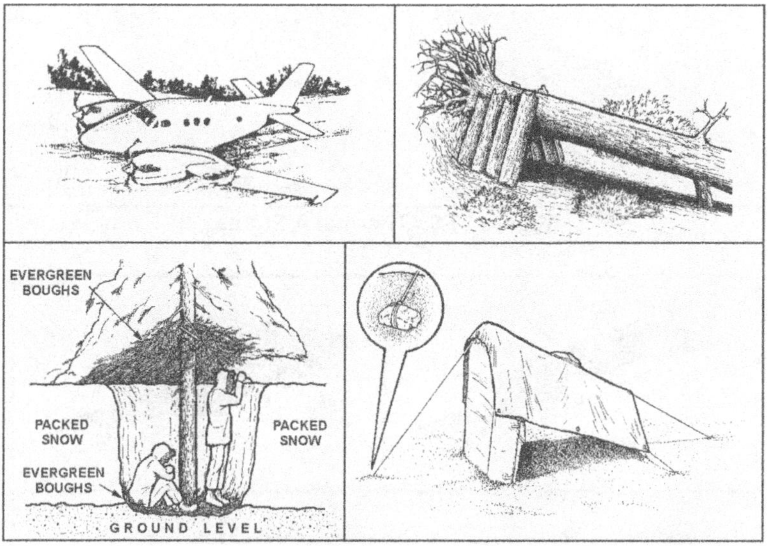


**Figure VI-4. Immediate Shelters**

(2) General shelter. Temperate climates require any shelter that gives protection from wind and rain.

(3) Thermal A Frame, Snow Trench, Snow Cave. **(Figures VI-5 through VI-7)**. Cold climates require an enclosed, insulated shelter.

(a) Snow is the most abundant insulating material.

(b) Air vent is required to prevent carbon monoxide poisoning when using an open flame inside enclosed shelters.

**Note:** As a general rule, unless you can see your breath, your snow shelter is too warm and should be cooled down to preclude melting and dripping.

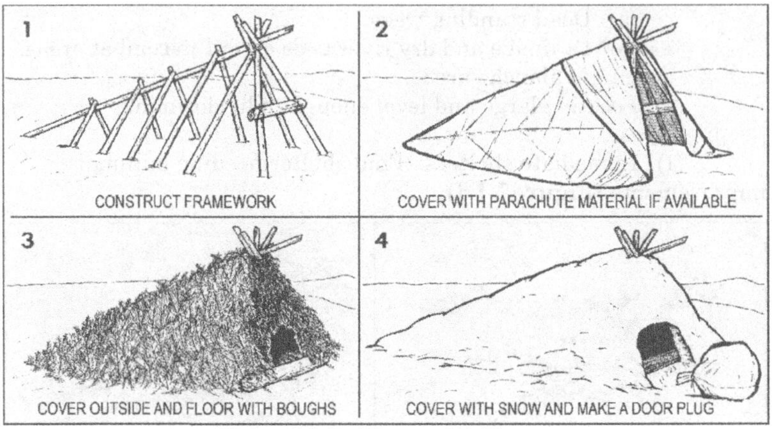


**Figure VI-5. Thermal A Frame**

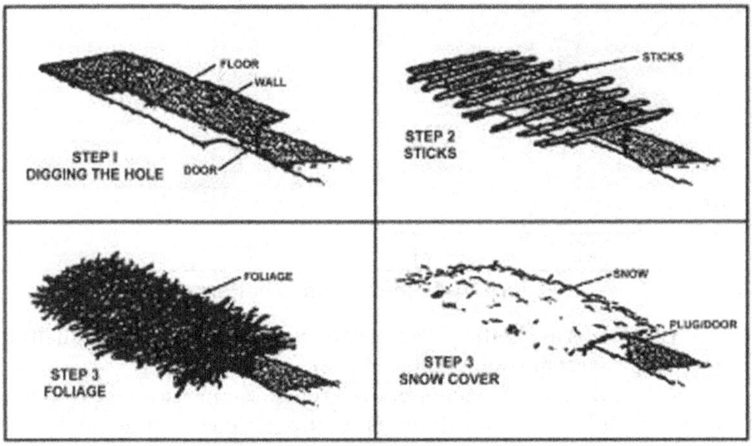


**Figure VI-6. Snow Trench**

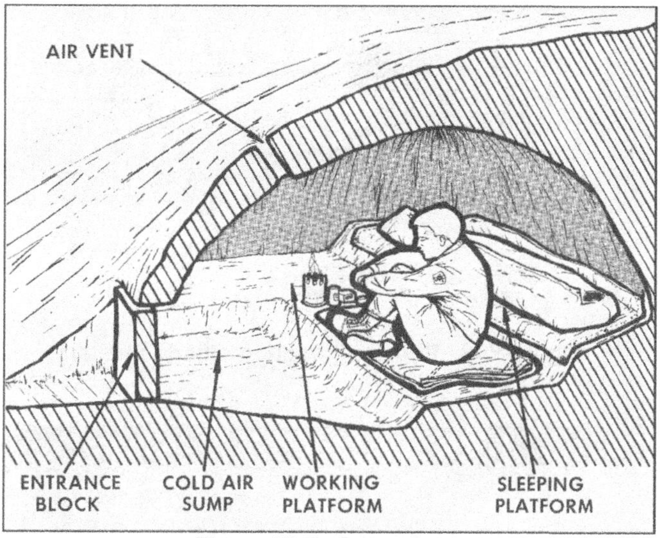


**Figure VI-7. Snow Cave**

(4) Shade shelter. Hot climates require a shade shelter to protect from ultraviolet rays **(Figure VI-8)**.

(a) To reduce the surface temperature, the shelter floor should be elevated or dug down (approximately 18 inches).

(b) For thermal protection, a minimum of **2** layers of material suspended 12-18 inches above the head is required. White is the best color to reflect heat (inner most layer should be of darker material).

(5) Elevated platform shelter **(Figure VI-9)**. Tropical/wet climates require enclosed, elevated shelter for protection from dampness and insects.

c. Shelter construction.

(1) Have entrance 45-90 degrees from prevailing wind.

(2) Cover with available material.

(a) If natural materials are used, arrange them in layers starting at the bottom with each layer overlapping the previous one. See **Figure VI-10** for an example.

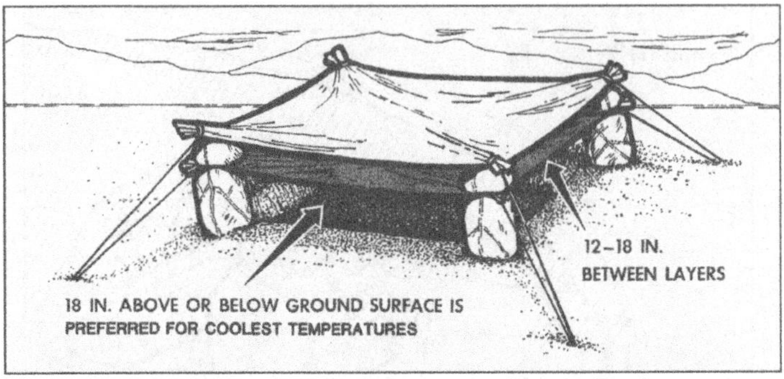


**Figure VI-8. Poncho/Parachute Shade Shelter**

**Figure VI-9. Elevated Platform Shelter**

**Figure VI-10. Shingle Method**

(b) If using porous material like parachute, blankets, etc.—

- Stretch as tight as possible
- Use a 40–60 degree slope.
- Use additional layers in heavy rain.

d. Shelter construction materials:

(1) Raft and raft parts.

(2) Vehicle or aircraft parts.

(3) Blankets, poncho, or parachute material.

(4) Sheet of plastic or plastic bag.

(5) Bark peeled off dead trees.

(6) Boughs, broad leaves, dry moss.

(7) Grass and sod.

(8) Snow.

(9) Sand and rocks.

e. Bed construction. Construct a bed to protect from cold, damp, ground using—

(1) Raft or foam rubber from vehicle seats.

(2) Boughs, leaves, or dry moss.

**5. Fires**

**CAUTION:** Weigh hazards and risks of detection against the need for a fire.

a. Evasion considerations:

(1) Use trees or other sources to dissipate smoke.

(2) Use fires at dusk, dawn, or during inclement weather.

(3) Use fires at times when the local populace is cooking.

b. Fire building. The **3** essential elements for starting a fire are heat, fuel, and oxygen.

(1) Heat sources:

(a) Matches or lighter.

(b) Flint and steel (experiment with various rocks and metals until a good spark is produced).

(c) Sparks from batteries.

(d) Concentrated sunlight (use magnifying glass or flashlight reflectors).

(e) Pyrotechnics, such as flares **(last resort)**, etc.

(f) Friction method **(Figure VI-11)**. Without prior training, this method is difficult to master and requires a lot of time to build the device.

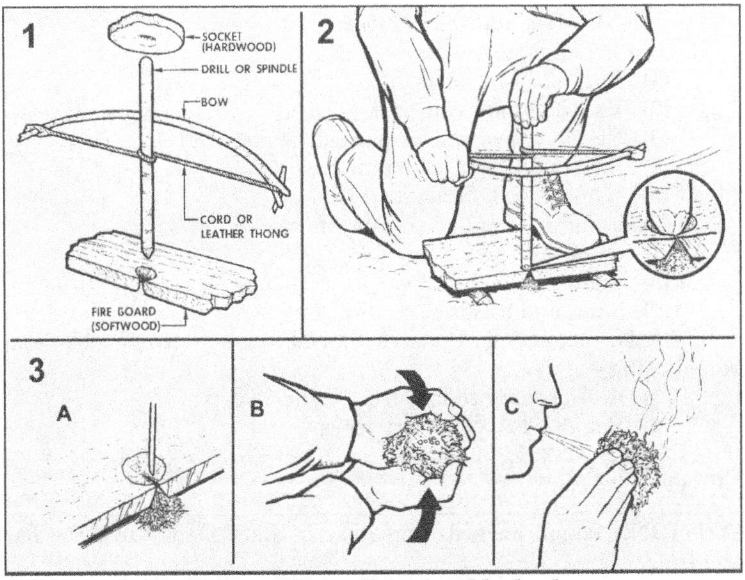


**Figure VI-11. Friction Method**

> **Note**: If possible, carry a fire-starting device with you.

(2) Fuel is divided into **3** categories: tinder, kindling, and fuel. (Gather large amounts of each category before igniting the fire.)

(a) **Tinder**. Tinder must be very finely shaved or shredded to provide a low combustion point and fluffed to allow oxygen to flow through. (To get tinder to burn hotter and longer, saturate with Vaseline, Chapstick, insect repellant, aircraft fuel, etc.) Examples of tinder include—

- Cotton.
- Candle (shred the wick, not the wax).
- Plastic spoon, fork, or knife.
- Foam rubber.

•Dry bark.
•Dry grasses.
•Gun powder.
•Pitch.
•Petroleum products.

(b) **Kindling**. Kindling must be small enough to ignite from the small flame of the tinder. Gradually add larger kindling until arriving at the size of fuel to burn.

(c) **Fuel**. Examples of fuel include—

•Dry hardwood (removing bark reduces smoke).
•Bamboo (open chambers to prevent explosion).
•Dry dung.

c. Types. Fires are built to meet specific needs or uses.

(1) Tepee fire **(Figure VI-12)**. Use the tepee fire to produce a concentrated heat source for cooking, lighting, or signaling.

**Figure VI-12. Tepee Fire**

(2) Log cabin fire **(Figure VI-13)**. Use the log cabin fire to produce large amounts of light and heat, to dry out wet wood, and provide coals for cooking, etc.

Figure VI-13. Log Cabin or Pyramid Fires

(3) Sod fire and reflector (**Figure VI-14**). Use fire reflectors to get the most warmth from a fire. Build fires against rocks or logs.

**CAUTION:** ***DO NOT*** use porous rocks or riverbed rock—they may explode when heated.

Figure VI-14. Sod Fire and Reflector

(4) Dakota fire hole (Figure VI-15). Use the Dakota fire hole for high winds or evasion situations.

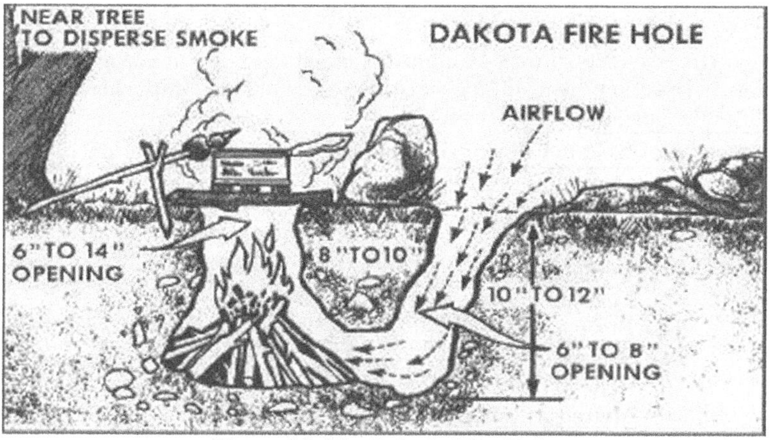


**Figure VI-15. Dakota Fire Hole**

(5) Improvised stoves **(Figure VI-16)**. These are very efficient.

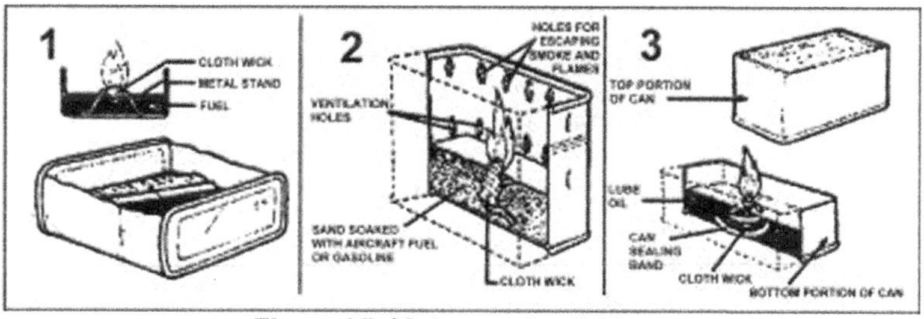


**Figure VI-16. Improvised Stove**

Chapter VII

# Water

## 1. Water Requirements

Drink **extra** water. Minimum 2 quarts per day to maintain fluid level. Exertion, heat, injury, or an illness increases water loss.

> **Note:** Pale yellow urine indicates adequate hydration.

## 2. Water Procurement

a. ***DO NOT*** drink—

(1) Urine.

(2) Fish juices.

(3) Blood.

(4) Sea water.

(5) Alcohol.

(6) Melted water from new sea ice.

b. Water sources:

(1) Surface water (streams, lakes, and springs).

(2) Precipitation (rain, snow, dew, sleet) **(FigureVII-1)**.

(3) Subsurface (wells and cisterns).

(4) Ground water (when no surface water is available) **(Figure VII-2)**.

(a) Abundance of lush green vegetation.

(b) Drainages and low-lying areas.

(c) *"V"* intersecting game trails often point to water.

(d) Presence of swarming insects indicates water is near.

(e) Bird flight in the early morning or late afternoon might indicate the direction to water.

(5) Snow or ice.

(a) ***DO NOT*** eat ice or snow.

- Lowers body temperature.
- Induces dehydration.
- Causes minor cold injury to lips and mouth.

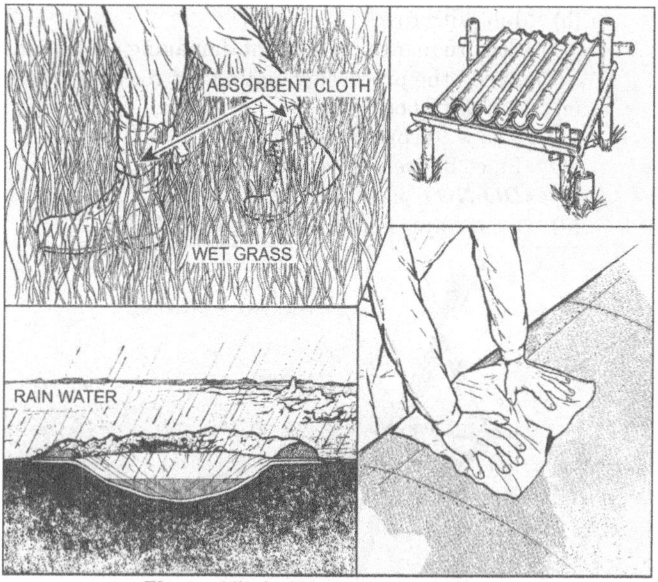


**Figure VII-1. Water Procurement**

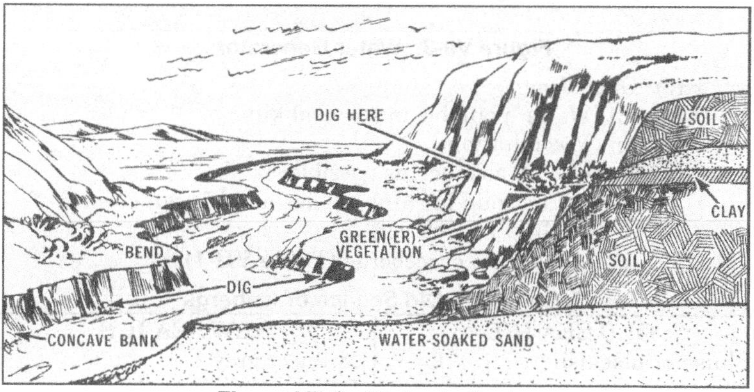


**Figure VII-2. Water Indicators**

(b) Melt with fire.
- Stir frequently to prevent damaging container.
- Speed the process by adding hot rocks or water.

(c) Melt with body heat.
- Use waterproof container.
- Place between layers of clothing.
- ***DO NOT*** place next to the skin.

(d) Use a water generator **(Figure VII-3)**.

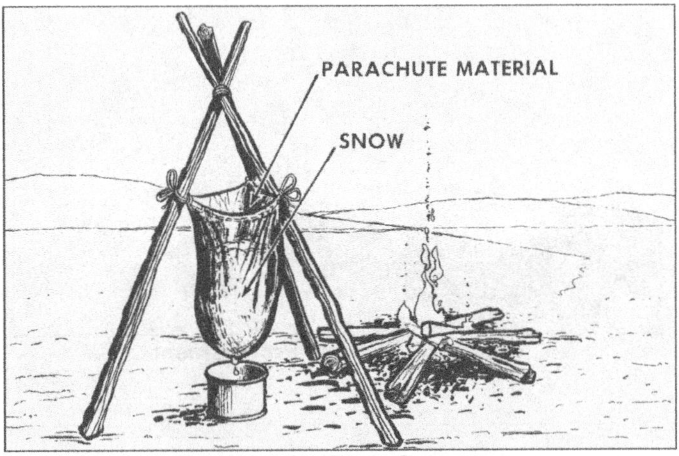


**Figure VII-3. Water Generator**

(6) Open seas.

(a) Water available in survival kits.

(b) Precipitation.
- Drink as much as possible.
- Catch rain in spray shields and life raft covers.
- Collect dew off raft.

(c) Old sea ice or icebergs **(Table VII-1)**.

**Table VII-1. Old Sea Ice or Icebergs**

| OLD SEA ICE | NEW SEA ICE |
|---|---|
| Bluish or blackish | Milky or grey |
| Shatters easily | Does not break easily |
| Rounded corners | Sharp edges |
| Tastes relatively salt-free | Tastes extremely salty |

(7) Tropical areas.

(a) All open sources previously mentioned.

(b) Vegetation.

•Plants with hollow sections can collect moisture.

•Leaning Tree. Cloth absorbs rain running down tree and drips into container **(Figure VII-4)**.

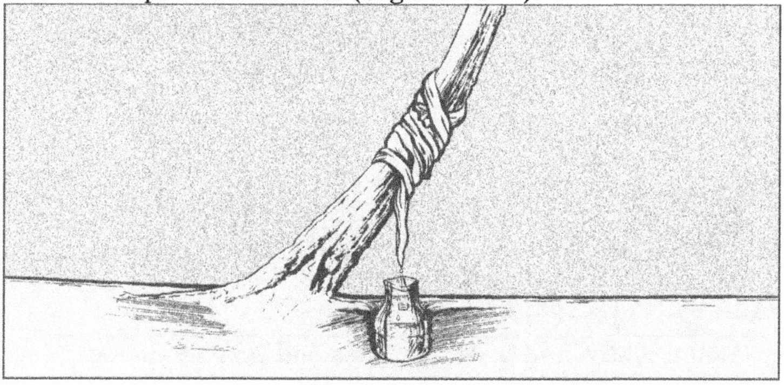

**Figure VII-4. Leaning Tree**

•Banana plants.

•Water trees (avoid milky sap).

••Tap before dark. Let sap stop running and harden during the daytime.

••Produce most water at night.

••For evasion situations, bore into the roots and collect water.

•Vines **(Figure VII-5A)**.

••Cut bark (***DO NOT*** use milky sap).

••If juice is clear and water like, cut as large a piece of vine as possible (cut the top first).

••Pour into hand to check smell, color, and taste to determine if drinkable.

••***DO NOT*** touch vine to lips.

••When water flow stops, cut off 6 inches of opposite end, water will flow again.

•Old bamboo.

••Shake and listen for water.

••Bore hole at bottom of section to obtain water.

••Cut out entire section to carry with you.

••Filter and purify.

•Green bamboo **(Figure VII-5B)**.

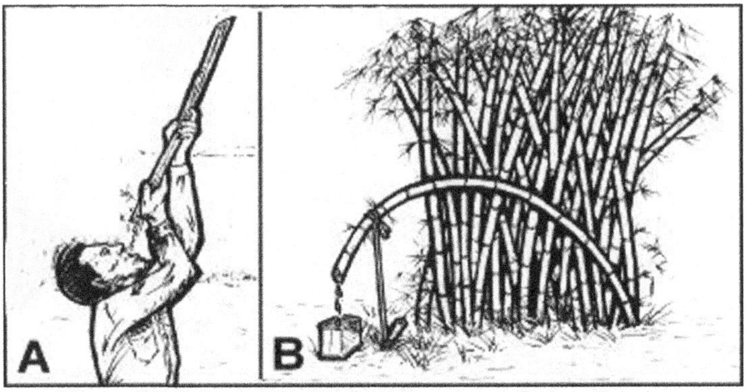


**Figure VII-5 A and B. Water Vines and Green Bamboo**

> **CAUTION:** Liquid contained in green coconuts (**ripe** coconuts may cause diarrhea).

•Beach well. Along the coast, obtain water by digging a beach well **(Figure VII-6)**.

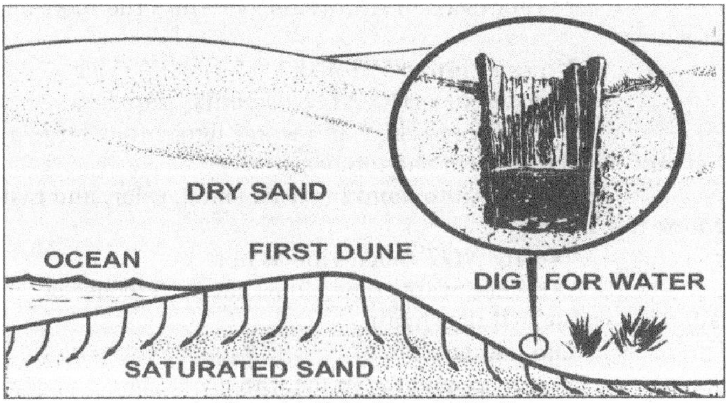


**Figure VII-6. Beach Well**

(8) Dry areas.

(a) Solar still **(Figure VII-7)**.

(b) Vegetation bag **(Figure VII-8)**.

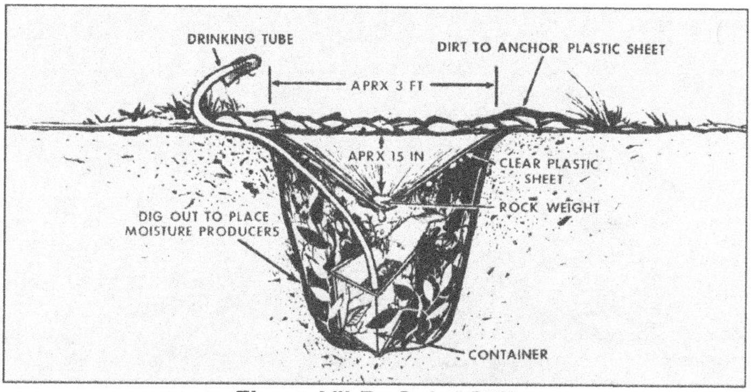


**Figure VII-7. Solar Still**

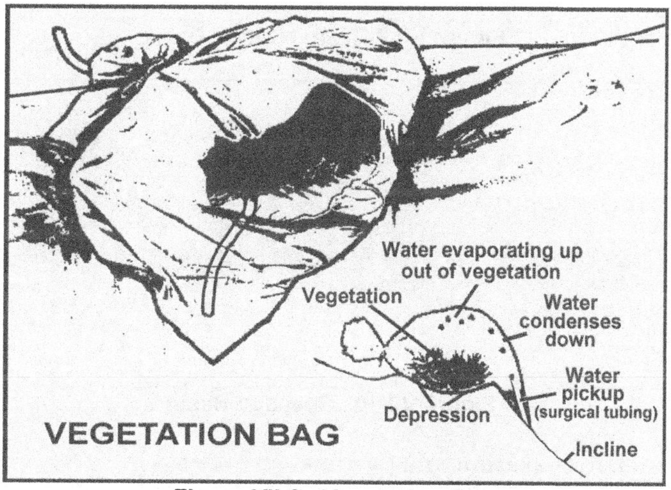


**Figure VII-8. Vegetation Bag**

(c) Transpiration bag **(Figure VII-9)**.

•Water bag must be clear.

•Water will taste like the plant smells.

(d) Seepage basin **(Figure VII-10)**.

> **CAUTION:** ***DO NOT*** use poisonous/toxic plants in vegetation/transpiration bags.

**Figure VII-9. Transpiration Bag**

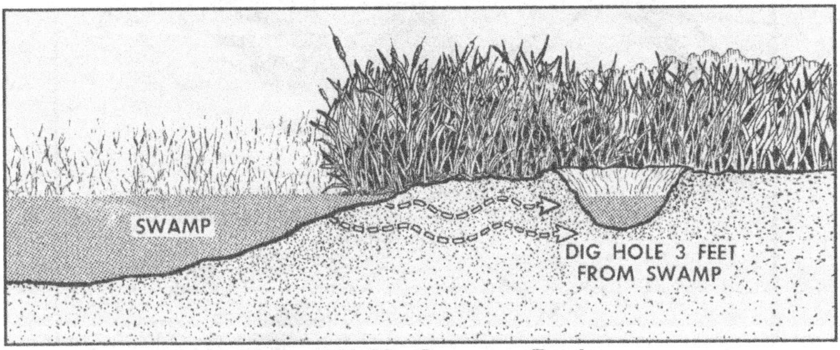


**Figure VII-10. Seepage Basin**

**3. Water Preparation and Storage**

a. Filtration. Filter through porous material (sand/charcoal).

b. Purification.

(1) Water from live plants requires no further treatment.

(2) Purify all other water.

(a) Boil at least 1 minute.

(b) Pour from one container to another to improve taste to aerate.

(c) Water purification tablets. Follow instructions on package.

c. Potable Water.

(1) If water cannot be purified, obtain water from a clear, cold, clean, and fast running source (if possible).

(2) Put in clear container and expose to the sun's ultraviolet rays to kill bacteria.

d. Storage. To prevent contamination, use a clean, covered or sealed container.

(1) Trash bag.

(2) Prophylactic.

(3) Section of bamboo.

(4) Flotation gear.

Chapter VIII

# FOOD

## 1. Food Procurement

a. Sources and location.

(1) Mammals can be found where—

(a) Trails lead to watering, feeding, and bedding areas.

(b) Droppings or tracks look fresh.

(2) Birds can be found by—

(a) Observing the direction of flight in the early morning and late afternoon (leads to feeding, watering, and roosting areas).

(b) Listening for bird noises (indication of nesting areas

(3) Fish and other marine life locations **(Figure VIII-1)**.

**Figure VIII-1. Fishing Locations**

(4) Reptiles and amphibians are found almost worldwide.

(5) Insects are found—

(a) In dead logs and stumps.

(b) At ant and termite mounds.

(c) On ponds, lakes, and slow moving streams.

b. Procurement techniques.

(1) Snares—

(a) Work while unattended.

(b) Location:

•Trails leading to water, feeding, and bedding areas.

•Mouth of dens **(Figure VIII-2)**.

**Figure VIII-2. Snare Placement**

(c) Construction of simple loop snare.

•Use materials that will not break under the strain of holding an animal.

•Use a figure 8 (locking loop) if wire is used **(Figure VIII-3)**.

••Once tightened, the wire locks in place, preventing reopening, and the animal's escape.

•To construct a squirrel pole **(Figure VIII-4)** use simple loop snares.

•Make noose opening slightly larger than the animal's head **(3-finger** width for squirrels, **fist-sized** for rabbits**)**.

(d) Placement of snares (set as many as possible).

•Avoid disturbing the area.

•Use funneling (natural or improvised) **(Figure VIII-5)**.

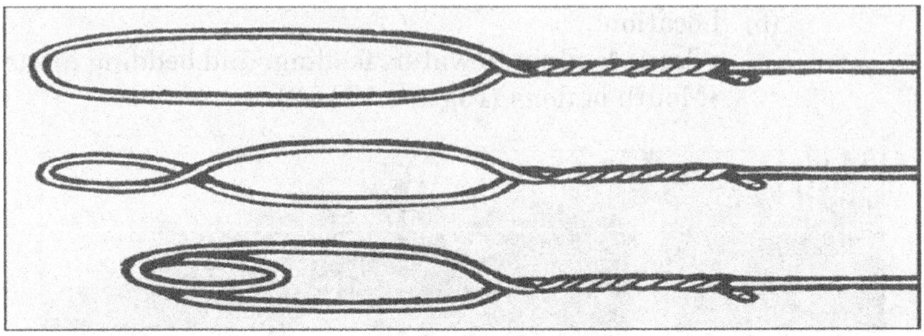

Figure VIII-3. Locking Loop

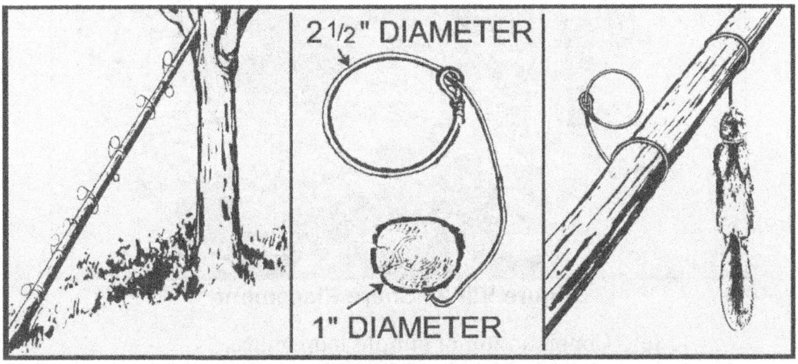


Figure VIII-4. Squirrel Pole

Figure VIII-5. Funneling

(2) Noose stick (easier and safer to use than the hands).

(3) Twist stick **(Figure VIII-6)**.

(a) Insert forked stick into a den until something soft is met.

(b) Twist the stick, binding the animal's hide in the fork.

(c) Remove the animal from the den.

(d) Be ready to **kill** the animal; **it may be dangerous**.

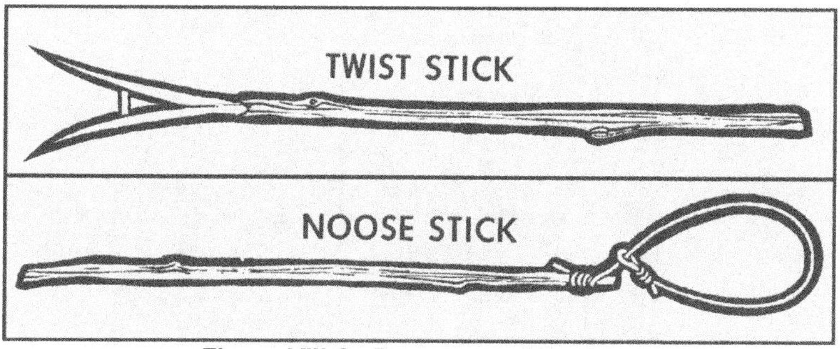


**Figure VIII-6. Procurement Devices**

(4) Hunting and fishing devices. (See **Figure VIII-7** for fishing procurement methods.)

(a) Club or rock.

(b) Spear.

(c) Slingshot.

(d) Pole, line, and hook.

(e) Net.

(f) Trap.

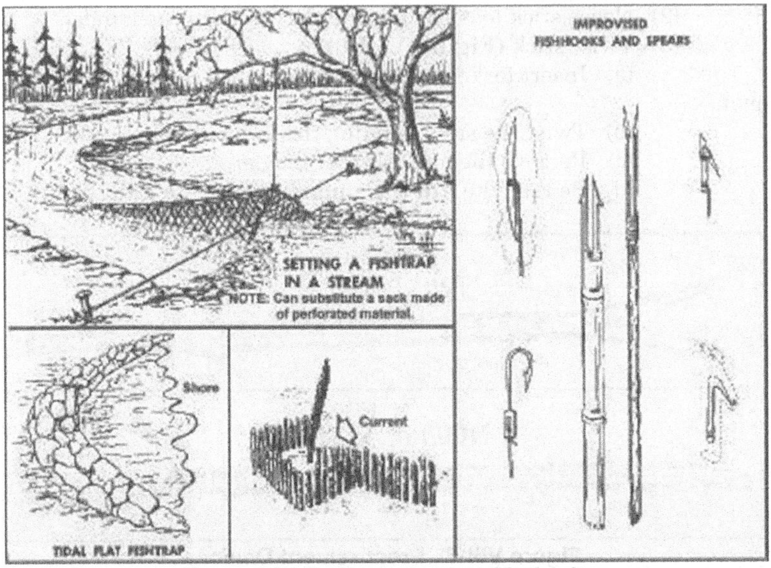


**Figure VIII-7. Procurement Methods**

(5) Precautions:

(a) Wear shoes to protect the feet when wading in water.

(b) Avoid reaching into dark holes.

(c) **Kill** animals before handling. Animals in distress may attract the enemy.

(d) ***DO NOT*** secure fishing lines to yourself or the raft.

(e) **Kill** fish before bringing them into the raft.

(f) ***DO NOT*** eat fish with—

•Spines.

•Unpleasant odor.

•Pale, slimy gills.

•Sunken eyes.

•Flabby skin.

•Flesh that remains dented when pressed.

(g) ***DO NOT*** eat fish eggs or liver (entrails).

(h) Avoid all crustaceans above the high tide mark.

(i) Avoid cone-shaped shells **(Figure VIII-8)**.

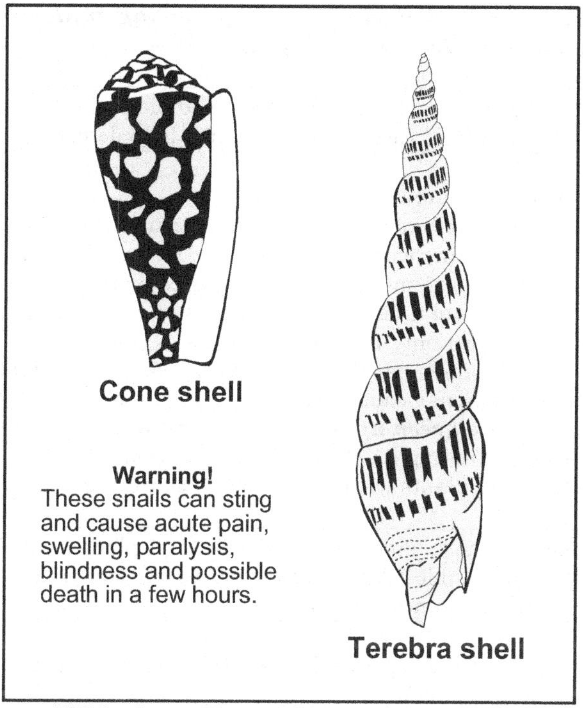


**Figure VIII-8. Cone-Shaped Shells of Venomous Snails**

(j) Avoid hairy insects; the hairs could cause irritation or infection.

(k) Avoid poisonous insects, for example:
- •Centipedes.
- •Scorpions.
- •Poisonous spiders.

(l) Avoid disease carrying insects, such as—
- •Flies.
- •Mosquitoes.
- •Ticks.

c. Plant Foods. ***Before using the following guide use your evasion chart to identify edible plants:***

> **Note**: If you cannot positively identify an edible plant and choose to try an unknown plant, these guidelines may help determine edibility.

(1) Selection criteria.

(a) Before testing for edibility, ensure there are enough plants to make testing worth your time and effort. Each part of a plant (roots, leaves, stems, bark, etc.) requires more than 24 hours to test. ***DO NOT*** waste time testing a plant that is not abundant.

(b) Test only **1** part of **1** plant at a time.

(c) Remember that eating large portions of plant food on an empty stomach may cause diarrhea, nausea, or cramps. ***Two*** good examples are ***green apples*** and ***wild onions***. Even after testing food and finding it safe, eat in moderation.

(2) Avoid plants with the following characteristics:

> **Note:** Using these guidelines in selecting plants for food may eliminate some edible plants; however, these guidelines will help prevent choosing potentially toxic plants.

(a) Milky sap (dandelion has milky sap but is safe to eat and easily recognizable).

(b) Spines, fine hairs, and thorns (skin irritants/contact dermatitis). ***Prickly pear*** and ***thistles*** are exceptions. ***Bracken fern fiddleheads*** also violate this guideline.

(c) Mushrooms and fungus.

(d) Umbrella shaped flowers (hemlock is eliminated).

(e) Bulbs (***only*** onions smell like onions).

(f) Grain heads with pink, purplish, or black spurs.

(g) Beans, bulbs, or seeds inside pods.

(h) Old or wilted leaves.

(i) Plants with shiny leaves.

(j) White and yellow berries. (Aggregate berries such as black and dewberries are always edible, test all others before eating.)

(k) Almond scent in woody parts and leaves.

d. Test procedures.

> **CAUTION:** Test all parts of the plant for edibility. Some plants have both edible and inedible parts. **NEVER ASSUME** a part that proved edible when cooked is edible raw, test the part raw before eating. The same part or plant may produce varying reactions in different individuals.

(1) Test only **1** part of a plant at a time.

(2) Separate the plant into its basic components (stems, roots, buds, and flowers).

(3) Smell the food for strong acid odors. Remember, smell alone does not indicate a plant is edible or inedible.

(4) ***DO NOT*** eat 8 hours before the test and drink only purified water.

(5) During the 8 hours you abstain from eating, test for contact poisoning by placing a piece of the plant on the inside of your elbow or wrist. The sap or juice should contact the skin. Usually 15 minutes is enough time to allow for a reaction.

(6) During testing, take ***NOTHING*** by mouth **EXCEPT** purified water and the plant you are testing.

(7) Select a small portion of a single part and prepare it the way you plan to eat it.

(8) Before placing the prepared plant in your mouth, touch a small portion (a pinch) to the outer surface of your lip to test for burning or itching.

(9) If after 3 minutes there is no reaction on your lip, place the plant on your tongue and hold it for 15 minutes.

(10) If there is no reaction, thoroughly chew a pinch and hold it in your mouth for 15 minutes **(*DO NOT* SWALLOW)**. If any ill effects occur, rinse out your mouth with water.

(11) If nothing abnormal occurs, swallow the food and wait 8 hours. If **any ill effects** occur during this period, **induce** vomiting and drink a water and charcoal mixture.

(12) If no ill effects occur, eat **¼ cup** of the same plant prepared the same way. Wait another 8 hours. If no ill effects occur, the plant part as prepared is safe for eating.

**CAUTION:**
1. Ripe tropical fruits should be peeled and eaten raw. Softness, rather than color, is the best indicator of ripeness. Cook unripe fruits and discard seeds and skin.
2. Cook underground portions when possible to reduce bacterial contamination and ease digestion of their generally high starch content.
3. During evasion, you may not be able to cook. Concentrate your efforts on leafy green plants, ripe fruits, and above ground ripe vegetables not requiring significant preparation.

**2. Food Preparation**

Animal food gives the greatest food value per pound.

a. Butchering and skinning.

(1) Mammals.

(a) Remove the skin and save for other uses.

(a) One cut skinning of small game **(Figure VIII-9)**.

•Open the abdominal cavity.

•Avoid rupturing the intestines.

•Remove the intestines.

•Save inner organs (heart, liver, and kidneys) and all meaty parts of the skull, brain, tongue, and eyes.

(b) Wash when ready to use.

(c) If preserving the meat, remove it from the bones.

(d) Unused or inedible organs and entrails may be used as bait for other game.

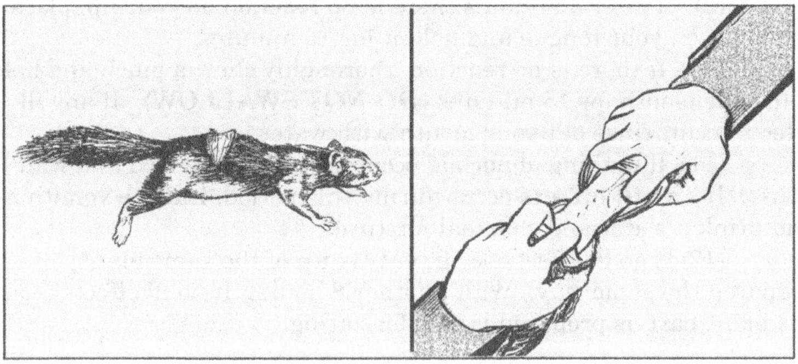

**Figure VIII-9. Small Game Skinning**

(2) Frogs and snakes.

(a) Skin.

(b) Discard skin, head with 2 inches of body, and internal organs.

(3) Fish.

(a) Scale (if necessary) and gut fish soon after it is caught.

(b) Insert knifepoint into anus of fish and cut open the belly.

(c) Remove entrails.

(d) Remove gills to prevent spoilage.

(4) Birds.

(a) Gut soon after killing.

(b) Protect from flies.

(c) Skin or pluck them.

(d) Skin scavengers and sea birds.

(5) Insects.

(a) Remove all hard portions such as the legs of grasshoppers or crickets. (The rest is edible.)

(b) Recommend cooking grasshopper-size insects.

**CAUTION:** Dead insects spoil rapidly, ***DO NOT*** save.

(6) Fruits, berries, and most nuts can be eaten raw.

b. Cooking.

**CAUTION:** To kill parasites, thoroughly cook all wild game, freshwater fish, clams, mussels, snails, crawfish, and scavenger birds. Saltwater fish may be eaten raw.

(1) Boiling (most nutritious method of cooking—drink the broth).

(a) Make metal cooking containers from ration cans.

(b) Drop heated rocks into containers to boil water or cook food.

(2) Baking.

(a) Wrap in leaves or pack in mud.

(b) Bury food in dirt under coals of fire.

(3) Leaching. Some nuts (acorns) must be leached to remove the bitter taste of tannin. Use one of the following leaching methods:

(a) First method:

•Soaking and pouring the water off.

•Crushing and pouring water through. Cold water should be tried first; however, boiling water is sometimes best.

•Discarding water.

(b) Second method:

•Boil, pour off water, and taste the plant.

•If bitter, repeat process until palatable.

(4) Roasting.

(a) Shake shelled nuts in a container with hot coals.

(b) Roast thinly sliced meat and insects over a candle.

**3. Food Preservation**

b. Keeping an animal alive.

c. Refrigerating.

(1) Long term.

(a) Food buried in snow maintains a temperature of approximately 32 degrees F.

(b) Frozen food will not decompose (freeze in meal-size portions).

(2) Short term.

(a) Food wrapped in waterproof material and placed in a stream remains cool in summer months.

(b) Earth below the surface, particularly in shady areas or along streams, is cooler than the surface.

(c) Wrap food in absorbent material such as cotton and re-wet as the water evaporates.

c. Drying and smoking removes moisture and preserves food.

(1) Use salt to improve flavor and promote drying.

(2) Cut or pound meat into thin strips.

(3) Remove fat.

(4) ***DO NOT*** use pitch woods such as fir or pine; they produce soot giving the meat an undesirable taste.

d. Protecting meat from animals and insects.
   (1) Wrapping food.
      (a) Use clean material.
      (b) Wrap pieces individually.
      (c) Ensure all corners of the wrapping are insect proof.
      (d) Wrap soft fruits and berries in leaves or moss.
   (2) Hanging meat.
      (a) Hang meat in the shade.
      (b) Cover during daylight hours to protect from insects.
   (3) Packing meat on the trail.
      (a) Wrap before flies appear in the morning.
      (b) Place meat in fabric or clothing for insulation.
      (c) Place meat inside the pack for carrying. Soft material acts as insulation helping keep the meat cool.
      (d) Carry shellfish, crabs, and shrimp in wet seaweed.

e. ***DO NOT*** store food in the shelter; it attracts unwanted animals.

# Chapter IX

# INDUCED CONDITIONS

## (NUCLEAR, BIOLOGICAL, AND CHEMICAL CONSIDERATIONS)

### 1. Nuclear Conditions

**CAUTION:** Radiation protection depends on time of exposure, distance from the source, and shielding.

a. Protection.

(1) **FIND PROTECTIVE SHELTER IMMEDIATELY!**

(2) Gather all equipment for survival (time permitting).

(3) Avoid detection and capture.

(a) Seek existing shelter that may be improved **(Figure IX-1)**.

Figure IX-1. Immediate Action Shelter

(b) If no shelter is available, dig a trench or foxhole as follows:

•Dig trench deep enough for protection, then enlarge for comfort **(Figure IX-2)**.

•Cover with available material.

**Figure IX-2. Improvised Shelter**

(4) Radiation shielding efficiencies **(Figure IX-3)**.

**NUCLEAR EXPLOSIONS:** Fall flat. Cover exposed body parts. Present minimal profile to direction of blast. ***DO NOT** look at fireball!* Remain prone until blast effects are over.

**SHELTER:** Pick, as soon as possible, 5 minutes unsheltered is maximum!

**Priority:**
(1) Cave or tunnel covered with 3 or more feet of earth.
(2) Storm/storage cellars
(3) Culverts.
(4) Basements.
(5) Abandoned stone/mud buildings.
(6) Foxhole 4 feet deep (remove topsoil within 2 feet radius of foxhole lip).

**RADIATION SHIELDING EFFICIENCIES**

| | | | |
|---|---|---|---|
| **Iron/Steel** | .7 inches | **Cinder Block** | 5.3 inches |
| **Brick** | 2.0 inches | **Ice** | 6.8 inches |
| **Concrete** | 2.2 inches | **Wood (Soft)** | 8.8 inches |
| **Earth** | 3.3 inches | **Snow** | 20.3 inches |

One thickness reduces received radiation dose by 1/2. Additional thickness added to any amount of thickness reduces received radiation dose by 1/2.

**SHELTER SURVIVAL:** Keep contaminated materials out of shelter.
**Good Weather:** Bury contaminated clothing outside of shelter (recover later).
**Bad Weather:** Shake strongly or beat with branches. Rinse and /or shake wet clothing. ***DO NOT** wring out!*

**PERSONAL HYGIENE:** Wash entire body with soap and *any* water; give close attention to fingernails and hairy parts.
**No Water:** Wipe all exposed skin surfaces with clean cloth or uncontaminated soil. Fallout/dusty conditions--keep entire body covered. Keep handkerchief/cloth over mouth and nose. Improvise goggles. ***DO NOT** smoke!*

**DAILY RADIATION TIME TABLE for NO RATE METER**

| | | | |
|---|---|---|---|
| **4-6** | Complete isolation | **9-12** | 2-4 hours exposure per day |
| **3-7** | Brief exposure (30 minutes maximum) | **13** | Normal movement. |
| **8** | Brief exposure (1 hour maximum) | | |

**Figure IX-3. Radiation Shielding Efficiencies**

(5) Leave contaminated equipment and clothing near shelter for retrieval after radioactive decay.

(6) Lie down, keep warm, sleep, and rest.

b. Substance:

(1) Water. Allow no more than 30 minutes exposure on **3d** day for water procurement.

(a) Water sources (in order of preference):

•Springs, wells, or underground sources are **safest**.

•Water in pipes/containers in abandoned buildings.

•Snow (**6** or **more inches below** the surface during the fallout).

•Streams and rivers (filtered before drinking).
•Lakes, ponds, pools, etc.
•Water from below the surface (***DO NOT*** stir up the water).
•Use a seep well.

(b) Water preparation **(Figures IX-4 and IX-5)**.
•Filtering through earth removes 99 percent of radioactivity.
•Purify all water sources.

(2) Food.

(a) Processed foods (canned or packaged) are preferred; wash and wipe containers before use.

(b) Animal foods.
•Avoid animals that appear to be sick or dying.
•Skin carefully to avoid contaminating the meat.
•Before cooking, cut meat away from the bone, leaving at least 1/8 inch of meat on the bone.
•Discard all internal organs.
•Cook all meat until **very well** done.

(c) Avoid.
•Aquatic food sources (use only in extreme emergencies because of high concentration of radiation).
•Shells of all eggs (contents will be safe to eat).
•Milk from animals.

(d) Plant foods (in order of preference).
•Plants whose edible portions grow underground (for example, potatoes, turnips, carrots, etc.). Wash and remove skin.
•Edible portions growing above ground that can be washed and peeled or skinned (bananas, apples, etc.).
•Smooth skinned vegetables, fruits, or above ground plants that are not easily peeled or washed.

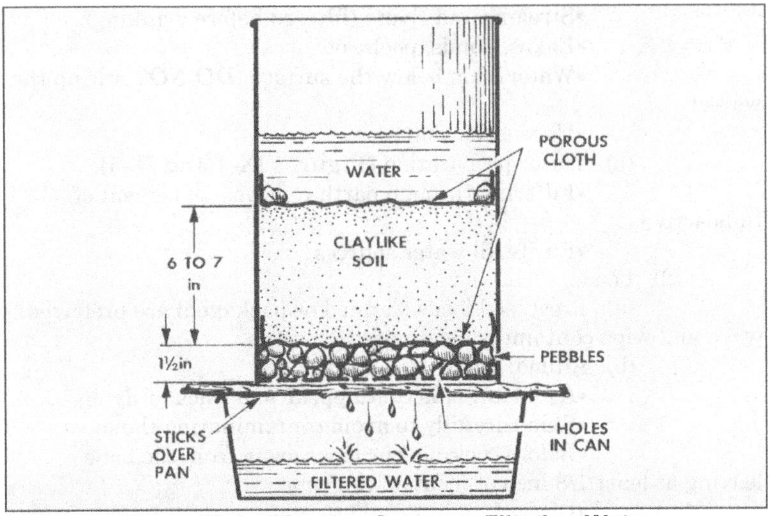


**Figure IX-4. Filtration Systems, Filtering Water**

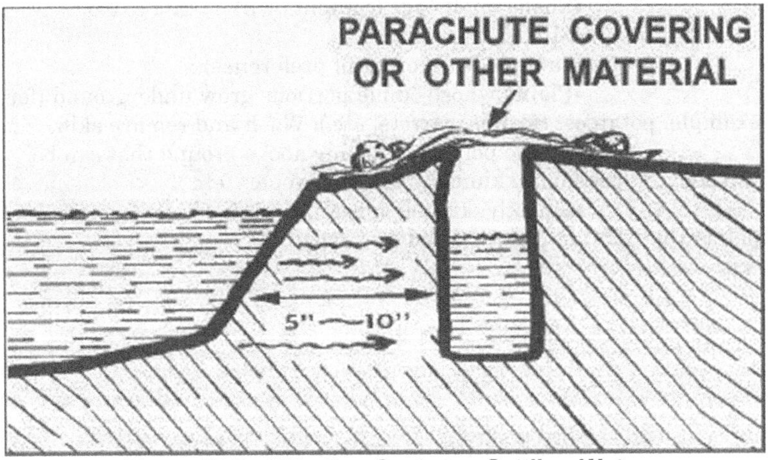


**Figure IX-5. Filtration Systems, Settling Water**

c. Self-aid:

(1) General rules:

(a) Prevent exposure to contaminants.

(b) Use personal hygiene practices and remove body waste from shelter.

(c) Rest, avoid fatigue.

(d) Drink liquids.

(2) Wounds.

(a) Clean affected area.

(b) Use antibacterial ointment or cleaning solution.

(c) Cover with clean dressing.

(d) Watch for signs of infection.

(3) Burns.

(a) Clean affected area.

(b) Cover with clean dressing.

(4) Radiation sickness (nausea, weakness, fatigue, vomiting, diarrhea, loss of hair, radiation burns).

(a) Time is required to overcome.

(b) Rest.

(c) Drink fluids.

(d) Maintain food intake.

(e) Prevent additional exposure.

**2. Biological Conditions**

a. Clues which may alert you to a biological attack follow:

(1) Enemy aircraft dropping objects or spraying.

(2) Breakable containers or unusual bombs, particularly those bursting with little or no blast, and muffled explosions.

(3) Smoke or mist of unknown origin.

(4) Unusual substances on the ground or vegetation; sick looking plants or crops.

b. Protection from biological agents follow:

(1) Use protective equipment.

(2) Bathe as soon as the situation permits.

(3) Wash hair and body thoroughly with soap and water.

(4) Clean thoroughly under fingernails.

(5) Clean teeth, gums, tongue, and roof of mouth frequently.

c. Survival tips for biological conditions follow:

(1) Keep your body and living area clean.

(2) Stay alert for clues of biological attack.

(3) Keep nose, mouth, and skin covered.

(4) Keep food and water protected. Bottled or canned foods are safe if sealed. If in doubt, boil food and water for 10 minutes.

(5) Construct shelter in a clear area, away from vegetation, with entrance 90 degrees to the prevailing wind.

(6) If traveling, travel crosswind or upwind (taking advantage of terrain to stay away from depressions).

**3. Chemical Conditions**

a. Detecting.

(1) Smell. Many agents have little or no odor.

(2) Sight. Many agents are colorless:

(a) Color. Yellow, orange, or red smoke or mist.

(b) Liquid. Oily, dark patches on leaves, ground, etc.

(c) Gas. Some agents appear as a mist immediately after shell burst.

(d) Solid. Most solid state agents have some color.

(3) Sound. Muffled explosions are possible indications of chemical agent bombs.

(4) Feel. Irritation to the nose, eyes, or skin and/or moisture on the skin are danger signs.

(5) Taste. Strange taste in food or water indicates contamination.

(6) General indications. Tears, difficult breathing, choking, itching, coughing, dizziness.

(7) Wildlife. Presence of sick or dying animals.

b. Protection against chemical agents follows:

(1) Use protective equipment.

(2) Avoid contaminated areas.

(a) Exit contaminated area by moving crosswind.

(b) Select routes on high ground.

(c) Avoid cellars, ditches, trenches, gullies, valleys, etc.

(d) Avoid woods, tall grasses, and bushes as they tend to hold chemical agent vapors.

(e) Decontaminate body and equipment as soon as possible by—

- Removing. Pinch-blotting.
- Neutralizing. Warm water.
- Destroying. Burying.

c. Self-aid in chemically contaminated areas.

(1) If a chemical defense ensemble is available—

(a) Use all protective equipment.

(b) Follow antidote directions when needed.

(2) If a chemical defense ensemble is not available—

(a) Remove or tear away contaminated clothing.

(b) Rinse contaminated areas with water.

(c) Improvise a breathing filter using materials available (T-shirt, handkerchief, fabric, etc.).

d. Tips for the survivor:

(1) ***DO NOT*** use wood from a contaminated area for fire.

(2) Look for signs of chemical agents around water sources before procurement (oil spots, foreign odors, dead fish, or animals).

(3) Keep food and water protected.

(4) ***DO NOT*** use plants for food or water in contaminated areas.

# Appendix A
# THE WILL TO SURVIVE

**ARTICLE VI CODE OF CONDUCT**

I will never forget that I am an American fighting for freedom, responsible for my actions, and dedicated to the principles which made my country free. I will trust in my God and in the United States of America.

**1. Psychology of Survival**

a. Preparation—

(1) Know your capabilities and limitations.

(2) Keep a positive attitude.

(3) Develop a realistic plan.

(4) Anticipate fears.

(5) Combat psychological stress by—

(a) Recognizing and anticipating existing *stressors* (injury, death, fatigue, illness, environment, hunger, isolation).

(b) Attributing normal reactions to existing *stressors* (fear, anxiety, guilt, boredom, depression, anger).

(c) Identifying signals of distress created by *stressors* (indecision, withdrawal, forgetfulness, carelessness, and propensity to make mistakes).

b. Strengthen your will to survive with—

(1) The Code of Conduct.

(2) Pledge of Allegiance.

(3) Faith in America.

(4) Patriotic songs.

(5) Thoughts of return to family and friends.

c. Group dynamics of survival include—

(1) Leadership, good organization, and cohesiveness promote high morale:

(a) Preventing panic.

(b) Creating strength and trust in one another.

(c) Favoring persistency in overcoming failure.

(d) Facilitating formulation of group goals.

(2) Taking care of your buddy.

(3) Working as a team.

(4) Reassuring and encouraging each other.

(5) Influencing factors are—
   (a) Enforcing the chain of command.
   (b) Organizing according to individual capabilities.
   (c) Accepting suggestions and criticism.

**2. Spiritual Considerations**

a. Collect your thoughts and emotions.
b. Identify your personal beliefs.
c. Use self-control.
d. Meditate.
e. Remember past inner sources to help you overcome adversity.
f. Pray for your God's help, strength, wisdom, and rescue.
   (1) Talk to your God.
   (2) Give thanks that God is with you.
   (3) Ask for God's help.
   (4) Pray for protection and a positive outcome.
g. Remember scripture, verses, or hymns; repeat them to yourself and to your God.
h. Worship without aid of written scripture, clergy, or others.
i. Forgive—
   (1) Yourself for what you have done or said that was wrong.
   (2) Those who have failed you.
j. Praise God and give thanks because—
   (1) God is bigger than your circumstances.
   (2) God will see you through (no matter what happens).
   (3) Hope comes from a belief in heaven and/or an after-life.
k. Trust.
   (1) Faith and trust in your God.
   (2) Love for family and self.
   (3) Never lose hope.
   (4) Never give up.
l. With other survivors—
   (1) Identify or appoint a religious lay leader.
   (2) Discuss what is important to you.
   (3) Share scriptures and songs.
   (4) Pray for each other.
   (5) Try to have worship services.

(6) Write down scriptures and songs that you remember.

(7) Encourage each other while waiting for rescue, remember—

(a) Your God loves you.

(b) Praise your God.

# Appendix B
# Publication Information

## 1. Scope

This UNCLASSIFIED multiservice tactics, techniques, and procedures publication is designed to assist Service members in a survival situation regardless of geographic location.

## 2. Purpose

This publication provides Service members a quick reference, weatherproof, pocket-sized guide on basic survival, evasion, and recovery information.

## 3. Application

The target audience for this publication is any Service member requiring basic survival, evasion, and recovery information.

## 4. Implementation Plan

Participating Service command offices of primary responsibility (OPRs) will review this publication, validate the information, and reference and incorporate it in Service and command manuals, regulations, and curricula as follows:

**Army.** The Army will incorporate the procedures in this publication in US Army training and doctrinal publications as directed by the commander, US Army Training and Doctrine Command (TRADOC). Distribution is in accordance with the Department of Army Form 12-99-R.

**Marine Corps.** The Marine Corps will incorporate the procedures in this publication in US Marine Corps training and doctrinal publications as directed by the commanding general, US Marine Corps Combat Development Command (MCCDC). Distribution is in accordance with Marine Corps Publication Distribution System.

**Navy.** The Navy will incorporate these procedures in US Navy doctrinal and training publications as directed by the commander, Navy Warfare Development Command (NWDC). Distribution is in accordance with Military Standard Requisitioning and Issue Procedures Desk Guide and Navy Standing Operating Procedures Publication 409.

**Air Force.** Air Force units will validate and incorporate appropriate procedures in accordance with applicable governing directives. Distribution is in accordance with Air Force Instructions 33-360.

**5. User Information**

a. The TRADOC-MCCDC-NWDC-AFDC Air Land Sea Application (ALSA) Center developed this publication with the joint participation of the approving Service commands. ALSA will review and update this publication as necessary.

b. This publication reflects current joint and Service doctrine, command and control (C2) organizations, facilities, personnel, responsibilities, and procedures. Changes in Service protocol, appropriately reflected in joint and Service publications, will likewise be incorporated in revisions to this document.

c. We encourage recommended changes for improving this publication. Key your comments to the specific page and paragraph and provide a rationale for each recommendation. Send comments and recommendation directly to—

## Army

Commander
US Army Training and Doctrine Command
ATTN: ATDO-A
Fort Monroe VA 23651-5000
DSN 680-3153 COMM (757) 727-3153

## Marine Corps

Commanding General
US Marine Corps Combat Development Command
ATTN: C42
3300 Russell Road
Quantico VA 22134-5001
DSN 278-6234 COMM (703) 784-6234

## Navy

Commander
Navy Warfare Development Command (Det Norfolk)
ATTN: ALSA Liaison Officer
1530 Gilbert Street
Norfolk VA 23511-2785
DSN 262-2782 COMM (757) 322-2782
E-mail: ndcjoint@nctamslant.navy.mil

## Air Force

HQ Air Force Doctrine Center
ATTN: DJ
216 Sweeney Boulevard Suite 109
Langley AFB VA 23665-2722
DSN 754-8091 COMM (757) 764-8091
E-mail Address: afdc.dj@langley.af.mil

## ALSA

ALSA Center
ATTN: Director
114 Andrews Street
Langley AFB, VA 23665-2785
DSN 575-0902 COMM (757) 225-0902
E-mail: alsadirector@langley.af.mil

*TC 21-24

TRAINING CIRCULAR
No. 21-24

HEADQUARTERS
DEPARTMENT OF THE ARMY
WASHINGTON, DC, 10 September 1997

# RAPPELLING

## CONTENTS



**DISTRIBUTION RESTRICTION:** Approved for public release; distribution is unlimited.

*This publication supersedes TC 21-24, 24 September 1991.

Page

## CHAPTER 3. HELICOPTER RAPPELLING



## CHAPTER 4. SPECIAL PATROL INFILTRATION/EXFILTRATION SYSTEM

# PREFACE

This circular provides basic rappelling techniques to soldiers and leaders for the conduct of rappelling operations. It serves as the primary reference for both resident and nonresident instruction presented to cadets, officer candidates, and both commissioned and noncommissioned officers. This circular also discusses several advanced techniques dealing with infiltration and exfiltration.

Safety is always the most important consideration when conducting training. This circular provides guidelines for commanders to conduct operations safely. Commanders at all levels must analyze the complete training event to determine the degree of risk involved to men and or equipment (Appendix A). After determining the risks, risk reduction options or controls will be integrated into the training activity. These options or controls may range from a safety briefing or providing additional safety resources, to selecting other means of accomplishing the mission.

Rappelling, FRIES, and SPIES operations are inherently dangerous; therefore, the safety notes and considerations presented in this circular are the minimum acceptable standards.

The proponent of this publication (except for Chapter 5) is the US Army Infantry School. Submit changes for improving this publication on DA Form 2028 and forward it to the Commandant, US Army Infantry School, ATTN: ATSH-RB, Fort Benning, Georgia 31905-5593 (e-mail address: IKNERH@BENNING-EMH2.ARMY.MIL).

The proponent for Chapter 5, Fast-Rope Insertion/Extraction System, is the US Army Special Operations Command. Submit changes or comments for Chapter 5 on DA Form 2028 and forward it to the Commander, US Army Special Operations Command, ATTN: AOOP-TRS, Fort Bragg, North Carolina 28307-5000.

Unless this publication states otherwise, masculine nouns and pronouns do not refer exclusively to men.

CHAPTER 1

# TOWER RAPPELLING

*The introduction to rappelling is taught on a static tower (Figure 1-1). Using the building block approach to training, soldiers systematically progress to more demanding platforms, including taller static towers, and finally to helicopters. The static tower used may vary in size and height from 34 to 90 feet. The concept of learning the basic rappelling techniques before helicopter operations does not vary. (Units may conduct ground training and wall-side tower rappelling without a rappel master. In this case, the unit commander would be responsible for the safety of the rappelling training.)*

**Figure 1-1. Rappel tower.**

**Note:** Each static rappel tower should resemble the structure or aircraft being trained to. The Army's standard design for building static rappel towers can be obtained through the Plans and Support Branch of the post Engineers, Building 75, Fort Benning, Georgia 31905 (commercial 706 545-3307 or DSN 835-3307). Strict adherence to the design specifications is required for safety considerations. Rappel towers should be inspected annually by post safety or post engineers.

## Section I. PERSONNEL

This section discusses the personnel involved in training rappelling and their duties and responsibilities.

### 1-1. RAPPEL MASTER

The proponent for accreditation, evaluation, and information for Army rappel master courses is Directorate of Operations and Training, US Army Infantry School, ATTN: ATSH-OTT-F, Fort Benning, Georgia 31905-5593.

a. **Duties and Responsibilities.** The rappel master is responsible for the safety of rappellers. He ensures that all equipment (installation, unit, and personal property) is serviceable. He personally supervises the rappelling operation.

b. **Qualification.** Rappel master qualification is awarded only after the successful completion of a TRADOC accredited rappel master course, which includes the following subjects:

- Duties and responsibilities of a rappel master.
- Safety SOP, regulations, and references.
- Construction of a deployment bag.
- Conduct of an equipment rappel off the rappel tower.
- Conduct of a lock-in.
- Talking a rappeller through completion of a rappel.
- Conduct of ground training.
- Inspecting for proper hook-ups.
- Inspecting and maintaining equipment.
- Inspecting and maintaining snaplinks.
- Inspecting and maintaining rappelling gloves.
- Inspecting and maintaining rappel ropes.
- Identifying the rappel capabilities of aircraft used.
- Controlling rappels from UH-1H or UH-60 aircraft.
- Tying knots (square, bowline, half-hitch, Prusik), safety lines, and rappel seats (Swiss seat, Australian seat).
- Inspecting a rappel seat.
- Aircraft rigging for rappelling operations.
- Aircraft command and control.

c. **Proficiency Maintenance.** To remain current, rappel masters must execute their duties in a tactical or training exercise once every six months. If rappel masters do not execute their duties once every six months, they must take a refresher class taught by a current rappel master. The refresher class includes the subjects listed in the rappel master qualifications paragraph (1-1b).

## 1-2. RAPPEL SAFETY OFFICER

The RSO is an air assault-qualified or ranger-qualified SFC or above who serves as the OIC for the conduct of rappel operations.

a. The RSO is responsible overall for the safety of all rappellers and ensures that all safety precautions are followed.

b. The RSO briefs VIPs, visitors, and inspecting authorities on training, safety requirements, and layout of training areas.

## 1-3. RAPPEL LANE NCO

a. **Duties and Responsibilities.** Safety is the rappel lane NCO's number one priority. The rappel lane NCO—

- Ensures proper safety procedures are followed.
- Ensures proper hookup once directed to a rope station.
- Issues commands and maintains eye contact with the rappeller at all times.

b. **Qualifications.** The rappel lane NCO holds the rank of corporal or above, is ranger- or air assault-qualified, and is selected by the commander. Each tower rappel lane must have a qualified rappel lane NCO.

c. **Training.** The rappel lane NCO must also train on the following subjects:

- Responsibilities and safety requirements.
- Inspection and maintenance of equipment.
- Identification of satisfactory anchor points.
- Identification of safe and unsafe hookups.
- Establishment of a rappel point.
- Inspection of a rappel seat.
- Coaching techniques.
- Rappelling procedures.
- Emergency procedures.
- Belay control procedures.

d. **Participation.** The rappel lane NCO must participate in at least seven rappel operations: three as a rappeller; two as an assistant rappel lane NCO; and two performing the duties of a rappel lane NCO under the supervision of a qualified rappel master.

e. **Proficiency Maintenance.** If a rappel lane NCO has not conducted his duties within the last six months, he must complete the training listed in paragraph 1-3c under the supervision of a current rappel master.

## 1-4. RAPPELLER

Rappel qualification requirements apply to the individual rappeller. Participants in tower rappel training must complete the following requirements under the supervision of a rappel master. The unit commander ensures that personnel successfully complete these requirements before beginning aircraft rappel training.

- Identify all rappelling equipment.
- Demonstrate the construction and attachment of the rappel seat and the rappel rope to the seat.
- Identify unsafe attachments, equipment, rope connections, and seat construction.
- Define terms used in rappelling operations.
- Identify knots used in rappel operations.
- Understand and demonstrate rappel commands.
- Demonstrate rappelling positions.
- Demonstrate belaying procedures.
- Exhibit satisfactory performance from a rappel tower of at least 34 feet in height (two rappels with equipment and weapon, two without equipment and weapon). Two rappels are conducted from the free side of the tower (no wall).
- Demonstrate the ability to lock in.

### 1-5. BELAYER

Belay requirements are a subtask of basic rappel requirements. Soldiers must know how to belay before conducting rappelling training. The belayer—

- Assumes a position at the base of the lane about one pace away from the tower area.
- Ensures that the rappel ropes are even with the ground during tower rappels.
- Loosely holds the rappel rope with both hands so as not to interfere with the rappeller but still be able to stop the rappeller should he fall.
- Immediately stops the rappeller by pulling downward on the rappel ropes if the rappeller shouts "falling" or loses control of his brake hand during descent.
- Does not wear gloves to ensure a firm grip on the rappelling rope.
- Watches the rappeller at all times, and maintains constant voice or visual contact.
- Wears a helmet to prevent injuries from falling debris.

### 1-6. BELAY SAFETY

The belay safety must be ranger- or air assault-qualified. He ensures belay personnel are performing their duties properly. Rappel training requires one belay safety for each two rappel stations.

## Section II. PREOPERATIONS BRIEFINGS AND SAFETY PROCEDURES

The rappel master ensures participants have a basic understanding of requirements and safety procedures before conducting training.

## 1-7. SAFETY

The following personnel and equipment must be present during static tower training.

- Two military rappel ropes for each rappel station.
- One safety officer.
- One rappel master for each rappel site.
- One rappel lane NCO per rappel station.
- One medic with medical kit and backboard.
- One safety or medical evacuation vehicle with driver.
- One belayer for each rope station. Rappellers alternate stations.
- One belay safety for each two rappel stations (four ropes).

## 1-8. SAFETY BRIEFING

As in all training, a safety briefing precedes rappel operations. The rappel master briefs all personnel on safety to include the following instructions.

a. Each rappeller ensures loose clothing and equipment are secured.

b. Rappel seats are tied by the soldier and inspected by the rappel master before climbing the tower. Rappel seats are removed upon completion of every rappel, retied, and reinspected by a qualified rappel master or rappel lane NCO before subsequent rappels.

c. Rappellers climb the tower only when directed by the rappel master or rappel lane NCO.

d. Rappellers stay in the center of the tower until instructed to move to a rappel point.

e. No more than three personnel are behind each rappelling point.

f. If using a troop ladder, only three soldiers are on the ladder at one time. Soldiers do not climb the ladder until told to do so by a rappel master.

g. All rappel masters, rappel lane NCOs, instructors, and anyone else standing near the edge of the top of the tower must wear a restraining strap or safety rope. The strap or rope must be attached to an anchor point.

h. No one should lean or sit on the railing or banisters of the tower.

i. No one is allowed within 3 feet of the edge of the tower without being secured.

j. When attaching the rappel rope to the snaplink, rappellers pull the slack towards the anchor point. The rappel master or rappel lane NCO physically check each hookup.

k. All personnel weighing more than 200 pounds will conduct a standard hookup rappel to determine if they require a friction hookup. A friction hookup is created by placing an additional two ropes in the gate of the snaplink (for a total of six ropes in the snaplink).

l. Combat equipment is positioned on the rappeller so that it does not interfere with the brake hand. The weapon must be slung diagonally across the back with the muzzle pointing down, and on the opposite side of the brake hand.

m. Heavy duty gloves are required for all rappel training.

n. Kevlar helmets with chin straps fastened are worn during tower rappel training.

o. While on the tower, the rappeller maintains eye contact with the rappel master or rappel lane NCO and receives all commands from them.

p. The rappeller ensures that he has a belayer on his rope.

q. The belayer does not wear gloves and keeps both hands on the rope at all times. He also faces the rappeller at all times.

r. All tower rappelling is performed with a double strand of rope.

s. No running is allowed on the tower.

t. No smoking or eating is allowed near the tower.

u. All participants who are unable to rappel, lack confidence, or refuse to rappel are reported to the rappel master or OIC. These participants are immediately removed from the training area.

v. The RSO and rappel master must be aware of overconfidence and carelessness of some rappellers. The rappel master ensures all personnel are tower-qualified before beginning aircraft rappel training.

## 1-9. TOWER SAFETY AND PREPARATION

The rappel master is in charge of the tower. He conducts a visual and physical inspection of every item of equipment to include the structural lumber and timber, the ladder, the platform floor, and all anchor points.

a. The static tower will not be used during thunderstorms or excessively high winds. If ice is present, or if the platform is slick from rain, rappelling will be delayed until conditions are safe.

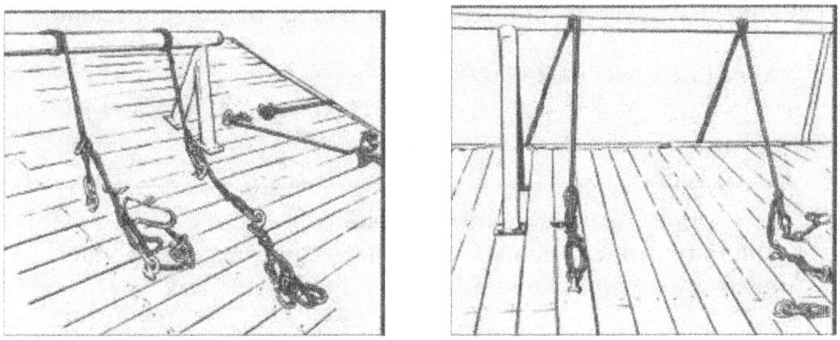

A B

**Figure 1-2. Example of rappel tower anchor points.**

b. All rope stations are rigged with two anchor points (Figure 1-2A, B, C). The first anchor point is a middle-of-the-rope knot, and the second is an end-of-the-rope anchor knot. The rappel master removes all the slack between the knots to create
equal tension on the anchor points. He ensures that no less than 10 feet of rope is on the ground during static rappelling.

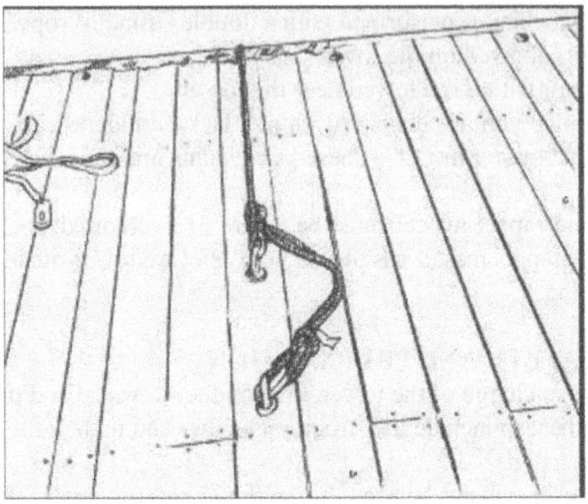

C

**Figure 1-2. Example of rappel tower anchor points (continued).**

## 1-10. RAPPELLER PREPARATION

Before conducting a rappel, each rappeller must prepare his individual clothing and equipment.

> **WARNING**
> **Failure to properly prepare rappellers could result in bodily injury or damage to equipment.**

a. Secure shirt tails, loose clothing, equipment, straps, and long hair.

b. Wear a helmet during rappelling. Properly fasten all straps and ensure the helmet is in serviceable condition.

c. Wear heavy leather workman's gloves.

d. Wear identification tags.

e. When rappelling with equipment, LBE or LBV should be unfastened in the front or fastened loosely behind the back of the rappeller. The rucksack should be worn high and tight on the back of the rappeller to allow the brake hand to reach the small of the back. Rucksack adjustment straps will be tied across the chest or tucked away.

f. Sling the weapon diagonally across the back with the muzzle down. Ensure the muzzle is on the guide hand side and the stock is towards the brake hand.

**Note:** Soldiers rappelling with equipment in excess of 50 pounds may want to consider using a friction hookup.

## Section III. RAPPELLING PROCEDURES

This section discusses procedures used in tower rappelling.

### 1-11. SEAT-HIP RAPPEL

When using the seat-hip rappel, friction is created by a snaplink that is inserted in a sling rope seat and fastened to the rappeller. This method provides a faster and more controlled descent than other methods. Wear gloves to prevent rope burns. An alternate technique is to insert a second snaplink into the first snaplink (attached to rappel seat) and run the rope through the second snaplink. This allows easier disengagement from the rappel rope without running the entire rope through the first snaplink. To disengage from the rappel rope using the alternate technique, release the tension from the rope by opening the gate of the first snaplink and removing the second snaplink (with the rope attached).

a. The rappel seat is constructed as follows (Figure 1-3A through 1-3T):

(1) Place the midpoint (center) of the length of the sling rope on the hip opposite the brake hand (the brake hand is the strong hand) (Figure 1-3A, B, C).

(2) Bring the sling rope around the waist above the hip bone. Tie a double overhand knot over the navel (Figure 1-3D, E, F, G, H).

(3) Let the two free ends of the sling rope fall to the ground in front (Figure 1-3I).

(4) Bring the two free ends of the sling rope down between the legs and up over the buttocks. Ensure that the two free ends do not cross (Figure 1-3J).

(5) Pass the ends of the ropes over the rope that is tied around the waist at the two points above the center of the two rear seat pockets (Figure 1-3K).

(6) Grab the free end of the rope that is on the left side of the body with the left hand, and the free end of the rope that is on the right side of the body with the right hand.

(7) Squat down and simultaneously pull on both running ends of the ropes and stand up. This will tighten the seat.

(8) Take the two running ends of the rope down and back over the waist rope from the inside. Bring the running ends back under the ropes that are going across the buttocks (Figure 1-3L).

(9) Tie the two running ends with a square knot and two overhand knots on the hip opposite the brake hand (Figure 1-3M, N).

(10) Place any excess rope in the trouser pocket near the square knot (Figure 1-3O, P).

(11) With the gate down and the hooked end of the snaplink against the navel, place the end of the snaplink through the single rope that is around the waist and the two ropes forming the double-hand knot (Figure 1-3Q).

(12) Rotate the snaplink a half turn so that the gate is facing up and will open away from the body (Figure 1-3R, S, T).

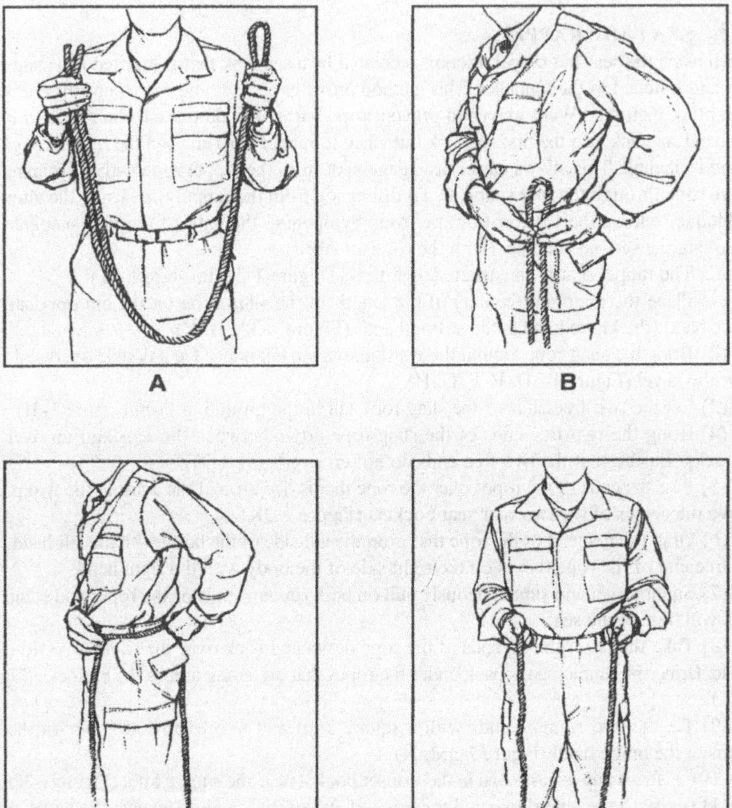


**Figure 1-3. Construction of rappel seat.**

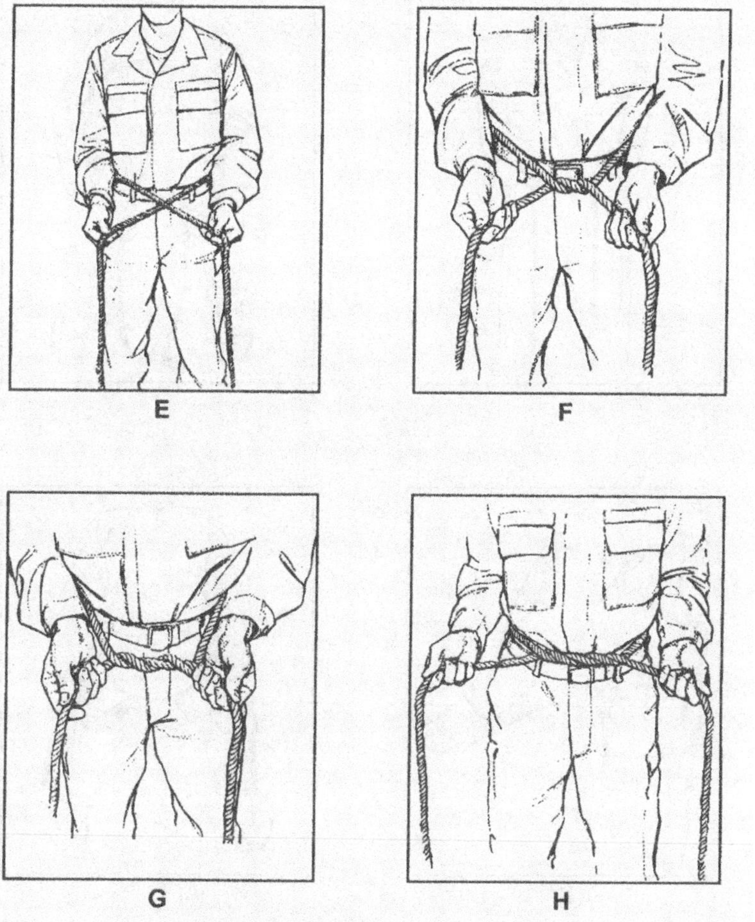


Figure 1-3. Construction of rappel seat.

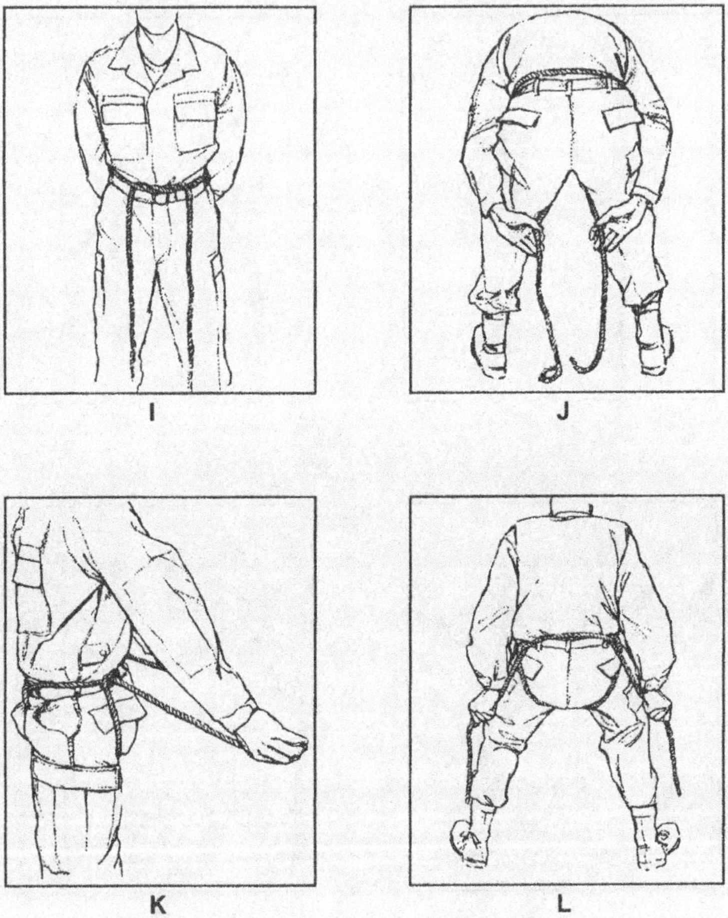

Figure 1-3. Construction of rappel seat.

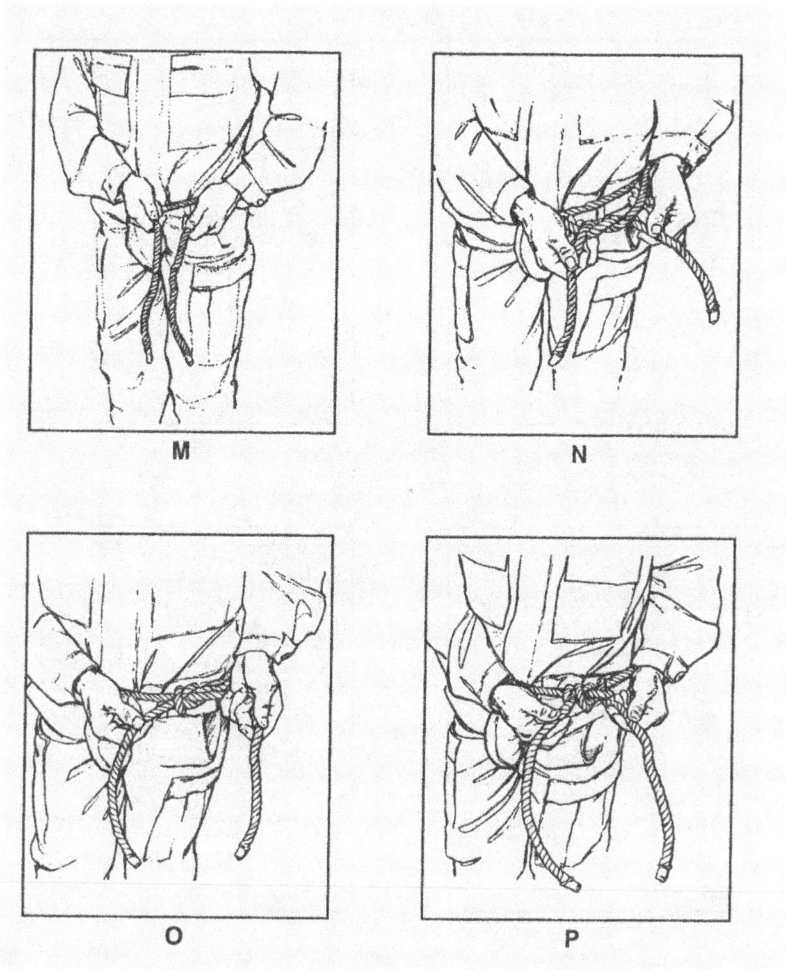


Figure 1-3. Construction of rappel seat.

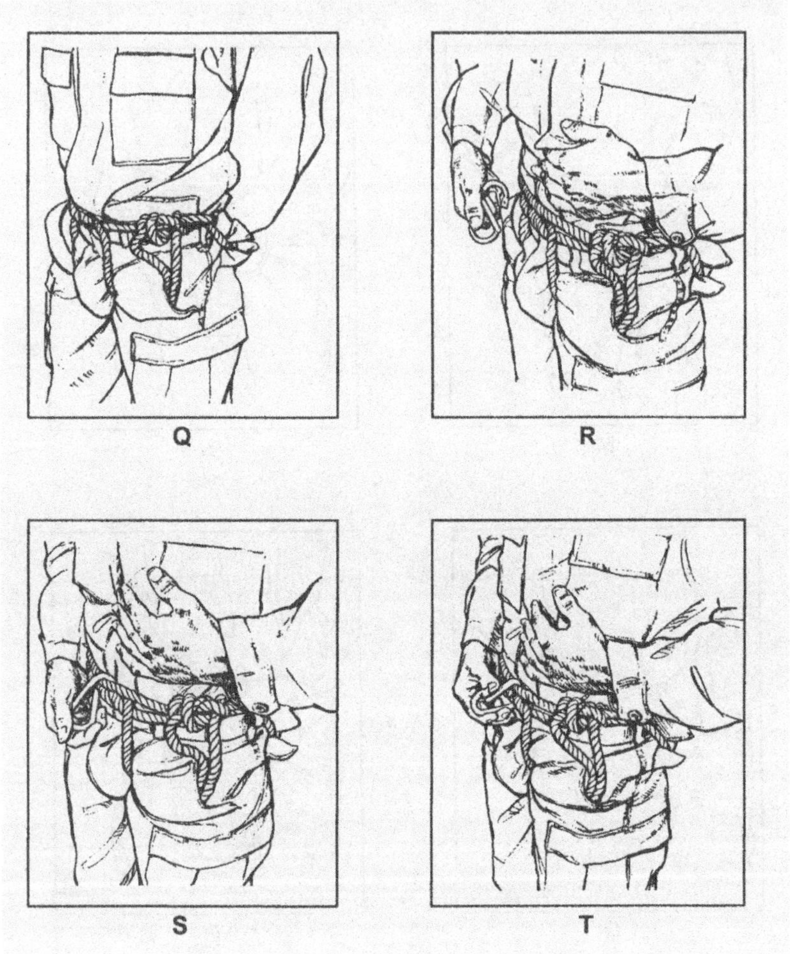


**Figure 1-3. Construction of rappel seat.**

b. To hook up using the seat-hip method, perform the following:

(1) Place the square knot with two overhand knots towards the anchor point for all seat-hip rappels.

(2) Grasp the two ropes with both hands and drop them through the gate of the snaplink. (At this point, two ropes should be running through the snaplink.)

(3) Using the hand closest to the anchor point, pull the slack towards the anchor point. Rotate the slack under and then over the top of the snaplink.

(4) Drop the two ropes a second time through the gate of the snaplink. (At this point, four ropes should be running through the snaplink.)

(5) Place the guide hand on the rope between the anchor point and the snaplink (palm facing up).

(6) Place the brake hand around the running end of the rope (palm facing down). Place the brake hand with the rope in the small of the back.

## 1-12. AUSTRALIAN RAPPEL

To hook up for the Australian rappel (Figure 1-4), perform the following:

a. Hook up to the Australian seat.

(1) Grasp the 9-foot sling rope at the midpoint (center) of its length, and double the rope.

(2) Place the doubled sling rope around the back and waist. Ensure that the rope is above the hipbone, but below the ribs.

(3) Tie a *square knot with two overhand knots* over the navel.

(4) Place any excess rope in the trouser pocket nearest the excess.

b. When the Australian seat is donned, face away from the anchor point and to the side of the rappel rope. (Stand to the same side of the rope as the brake hand. Determine left and right of the rope while facing the anchor point.) Place a snaplink onto the seat with the gate facing up, hinge closest to the body. Place it on the hip corresponding to the brake hand. The rappel master grasps the rappel rope and lays the rope into the snaplink. He then places one hand between the snaplink and the anchor point and draws slack toward the anchor point. He rotates the slack down, under, and over the rope and into the snaplink. The rappel master then slides the snaplink directly to the rappellers back.

c. Serve the running end of the rappel rope with the brake hand and prepare to rappel. During descent, brake by drawing the rappel ropes diagonally across the chest with the ropes running from near the waist to the pocket of the opposite shoulder.

**Figure 1-4. Australian rappel.**

## 1-13. CLIMBING PROCEDURES

Before climbing the ladder, the rappel master, safety OIC, or rappel lane NCO checks each rappeller's equipment.

a. The rappeller kicks the sand off his boots before climbing.

b. The rappeller grasps the outside of the ladder while climbing, not the rungs (when possible). If stairs are built for the tower, the rappeller grasps the railings as appropriate.

c. Just before climbing up the ladder, the rappeller sounds off, "(name) climbing," and then begins climbing up the ladder. Once at the top and clear of the ladder, the rappeller sounds off, "(name) clear."

d. Once off the ladder, the rappeller waits until the rappel master or lane NCO directs him to proceed to a rope station. At this time, the next rappeller in line may start to climb the ladder.

## 1-14. TOWER PROCEDURES

After the rappellers climb the tower, the following procedures are adhered to:

a. Once directed to a rope station, the rappel master or lane NCO ensures proper hookup for rappelling.

b. At this time, the rappeller sounds off with "on rappel" and the belayer sounds off with "on belay."

c. While maintaining his brake, the rappeller (on command from the rappel master or lane NCO) steps over the safety rail and faces the anchor point.

d. At this point, the rappel lane NCO sounds off with the following verbal commands and arm-and-hand signals.

(1) *Get Ready*—Rappel master or lane NCO extends both arms to the front with fists clenched and thumbs pointing upward (Figure 1-5). This alerts the rappeller. Each rappeller then looks over his brake hand shoulder to check for the belay man. The rappeller then looks at the rappel master. The rappel master or lane NCO makes his second check of the hookup, rappel seat, snaplink, and equipment.

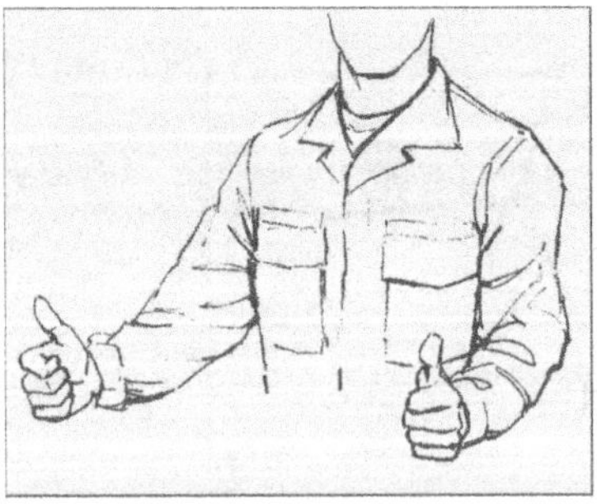

**Figure 1-5. Example of arm-and-hand signal GET READY.**

(2) *Position*—The lane NCO extends both arms to the front, elbows locked, forearms pointed downward, and fingers extended. He makes a circular motion with both forearms rotating in opposite directions (Figure 1-6). With the brake hand in the small of the back, the rappeller rotates 180 degrees out onto the wall or skid mock up and assumes an L-shaped position. The feet should be shoulder width apart, balls of feet on the wall or skid, knees locked, and body bent at the waist (Figure 1-7).

**Figure 1-6. Example of arm and hand signal POSITION.**

**Figure 1-7. L-shape position.**

(3) *Go*—The lane NCO extends his right arm with the elbow locked, fingers extended, thumbs around the index fingers, and points directly at the rappeller (Figure 1 8). This initiates the rappel. The rappeller flexes his knees and jumps vigorously backwards. At the same time, the rappeller throws his brake hand out at a 45-degree angle, letting the rope slide through both the brake hand and the guide hand. The rappeller looks over his brake hand shoulder at all times during descent.

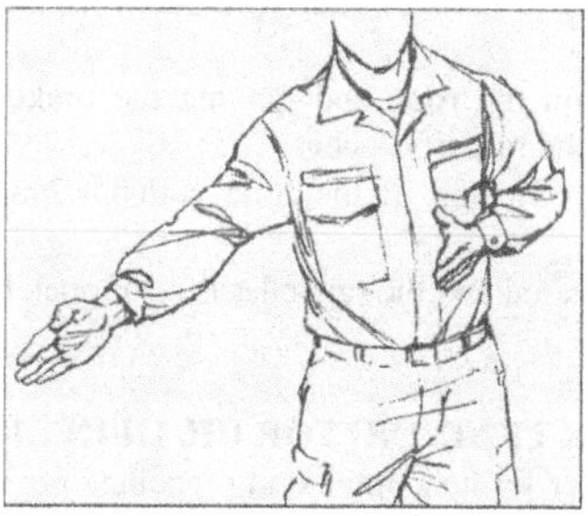

**Figure 1-8. Example of arm-and-hand signal GO**

e. The rappeller descends in a smooth, controlled manner.

f. The rappeller maintains eye contact with the ground at all times.

g. The rappeller maintains a modified L-shape position during descent with the feet shoulder-width apart, knees flexed, and buttocks parallel to the ground (Figure 1-9).

**Figure 1-9. L-shape position while rappelling.**

h. When carrying equipment or additional weight, a modified L-shape is used with the legs slightly lower than the buttocks to compensate for the added weight.

i. The rappeller's back is straight. He looks over the brake hand shoulder.

j. The guide hand is extended on the rope with the elbow extended and locked.

k. The rope slides freely through the guide hand, which is used to adjust equipment and to assist in balance during descent.

l. To brake, the rappeller places the brake hand (with rope in hand) in the small of the back and then grasps the rope firmly with the brake hand.

**Note:** Do not grip the rope firmly with the brake hand while the brake hand and brake arm are extended at the 45-degree angle. If this is done while rappelling, the brake hand and glove may become entangled in the snaplink causing injury to the hand and causing the rappeller to become hung up on the ropes.

m. Releasing tension on the rope and moving the brake hand out to the rear at a 45-degree angle regulates the rate of descent.

n. The rappeller never lets go of the rope with his brake hand until the rappel is completed.

o. After the rope is cleared and the rappeller is off rappel, he acts as the belayer for the next rappeller.

## 1-15. RAPPEL TOWER TRAINING FOR UH-1H HELICOPTER

Training on the rappel tower for helicopter skid rappelling prepares soldiers to rappel from a UH-1H helicopter (Figure 1-10). The rappeller is hooked up while he sits on the platform just above the helicopter skid. On the rappel master's command GET READY, the rappeller looks over the edge of the tower to ensure the running ends of the ropes are on the ground (Figure 1-10A). On the command SIT IN THE DOOR, the rappeller rotates his feet and legs off the platform and places them on the skid (Figure 1-10B). On the command POSITION, the rappeller turns around and assumes an L-shape position (Figure 1-10C). On the command GO, the rappeller bounds away from the helicopter skid and rappels to the ground. The rappel master is responsible for the proper procedures and safety.

A B C

**Figure 1-10. UH-1H helicopter skid rappel training.**

## 1-16. RAPPEL TOWER TRAINING FOR UH-60 BLACKHAWK HELICOPTER

Rappel tower training for the UH-60 Blackhawk is similar to that for the UH-1H with the exception that the UH-60 has no skids on which to stand. Therefore, the edge of the rappel tower is used as a pivot point to assume the L-shape position. All commands are the same except for SIT IN THE DOOR, which does not apply to the UH-60. The stances for each command are also different, as shown in Figure 1-11.

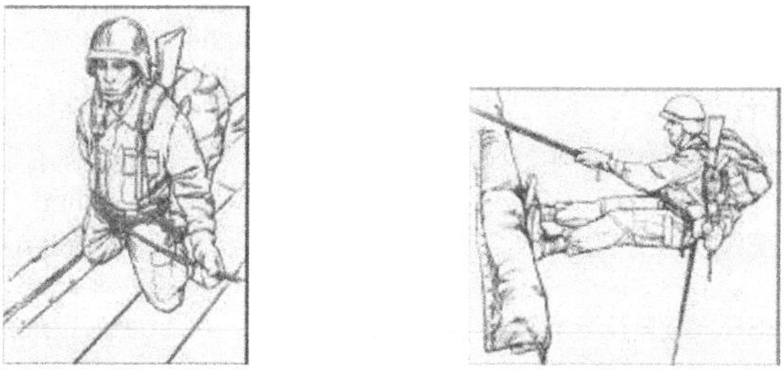

A B

**Figure 1-11. UH-60 helicopter rappel training.**

## 1-17. EMERGENCY LOCK-IN PROCEDURES

Mastering the lock-in procedure during tower training is critical before advancing to helicopter rappelling. Using the lock-in procedure allows personnel to hold in position for an extended period of time.

a. If the helicopter gains altitude above the length of the rappel ropes, the rappeller immediately brakes, locks in, and waits for the descent of the aircraft. Procedures for lock-in are as follows:

(1) Place the brake hand in the small of the back and brake to a complete stop.

(2) Release the guide hand from the ropes.

(3) Bring the guide hand around the back and grasp the running end of the two rappel ropes behind the brake hand.

(4) Using the guide hand, bring the two running ends of the rappel ropes around to the front.

(5) Secure these two running ends of the rappel ropes with the two anchor ends of the rappel ropes in the guide hand. This is now the new brake hand.

(6) Take the old brake hand out of the small of the back. Bring it around to the front and grasp the two ropes from the anchor point at a point just above the new brake hand. The old brake hand is now the new guide hand.

(7) Face the rappel master and wait for his command to lower to the ground.

(8) When the command is received from the rappel master to continue the descent, bring the brake hand to a 45-degree angle to the rear. When it is time to brake, bring the new brake hand around to the front diagonally across the chest.

b. If an engine fails or an aircraft emergency occurs during rappelling, the rappellers on the ropes descend as rapidly as possible and move from beneath the aircraft to the sides. Rappellers will maintain control of ropes, if possible.

## 1-18. COMMUNICATIONS

The rappeller at the top of a rappel point must be able to communicate with those at the bottom. Radios, hand signals, and rope signals are considered during a tactical rappel. For training situations, use the following commands:

| COMMAND | GIVEN BY | MEANING |
|---|---|---|
| LANE NUMBER, ON RAPPEL | Rappeller | I am ready to begin rappelling. |
| LANE NUMBER, ON BELAY | Belayer | I am on belay and you may begin your rappel. |
| LANE NUMBER, OFF RAPPEL | Rappeller | I have completed the rappel, cleared the rappel lane, and am off the rope. |
| LANE NUMBER, OFF BELAY | Belayer | I am off belay. |
| LANE NUMBER, FALLING | Rappeller | I am falling. Be alert below—belay man brake. |

## 1-18. DEMONSTRATION

After explaining the procedures to all rappellers, the rappel master should have an assistant demonstrate one complete cycle of rappelling from the static tower. This ensures that the rappellers can hear all the proper commands and see the actions and techniques used on the static tower.

CHAPTER 2

# GROUND RAPPELLING

*Ground rappelling is a technique that allows soldiers to negotiate mountains and cliffs safely and rapidly. Before rappellers participate in mountain and cliff rappelling they should have completed all the requirements for tower training. To ensure the rappel is safely conducted and meets the mission requirements, rappellers should consider several factors.*

## 2-1. PERSONNEL

The personnel responsibilities for mountain and cliff rappelling are the same as for tower rappelling. As a minimum the following personnel are needed for mountain and cliff rappelling.

a. **Mountaineering Safety Officer (MSO).** The MSO is a SFC or above who is ranger qualified or is a graduate of the Northern Warfare Training Center (NWTC). He has overall responsibility for the safety of all mountaineering participants and ensures that all safety precautions are followed.

b. **Rappel Master**. The rappel master is a qualified officer or NCO in charge of training on ground, tower, and aircraft rappel safety procedures.

c. **Medical Officer or Noncommissioned Officer.** Qualified rappellers and aidmen must know MEDEVAC procedures.

d. **Radiotelephone Operator.** The RATELO monitors the radio and reports to the RSO.

e. **Rappel Lane NCO.** The rappel lane NCO is a qualified officer or NCO in charge of rappelling on a wall or ground rappel lane (Chapter 1, paragraph 1-3).

## 2-2. SUSTAINMENT TRAINING

Before conducting ground rappel training, the unit or element conducts sustainment training.

a. Review the construction of a rappel seat, seats to be used, and hook-up procedures.

b. Conduct two rappels on the 34-foot tower wall (20-foot tower is acceptable): one with equipment, one without equipment. Conduct two rappels from the free side of a 34-foot or higher rappel tower.

c. Demonstrate the ability to lock in during one of the rappels from the free side of the tower.

## 2-3. SELECTION OF A RAPPEL POINT

The selection of the rappel point depends on factors such as mission, cover, route, anchor points, and edge composition (loose or jagged rocks). The anchor point should be above the rappeller's departure point. Primary and secondary anchors must be solid (natural anchors are preferred). The rappeller should be sure that the rope reaches the bottom or a place from which he can further rappel or climb. Also, the rappel point should be carefully tested and inspected to ensure the rope will run freely and that the area is clear of obstacles that could be dislodged. If a sling or runner is used for a rappel point, it should be tied twice from two separate loops. Suitable loading and off-loading platforms should be available.

## 2-4. ESTABLISHMENT OF A RAPPEL POINT

A rappel lane is the area and rope that a rappeller will use during his descent. It should have equal tension between all anchor points by establishing primary and secondary anchor points (Figure 2-1). If one anchor point fails, the rappel rope should not extend. All of the methods discussed below can be performed with a single or double rope. A double rope application should be used for safety.

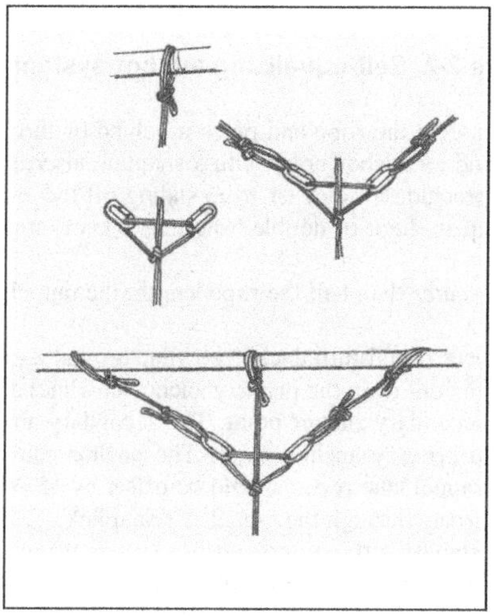

**Figure 2-1. Rappel point.**

a. If a rappel lane is *less* than half the rope length, the rappeller may apply one of the following techniques:

(1) Double the rope and tie a three-loop bowline around the primary anchor to include the primary anchor inside two loops and enough rope in the third loop to run to the secondary anchor (another three-loop bowline secured with an overhand knot).

(2) Double the rope and tie it around a secure anchor point with a round turn anchor bowline secured with an overhand knot (or any appropriate anchor knot).

(3) Double the rope and establish a self-equalizing anchor system (Figure 2-2) with a bowline-on-a-bight or figure-eight-on-a-bight knot. Tie off on the long standing and with a round turn anchor bowline.

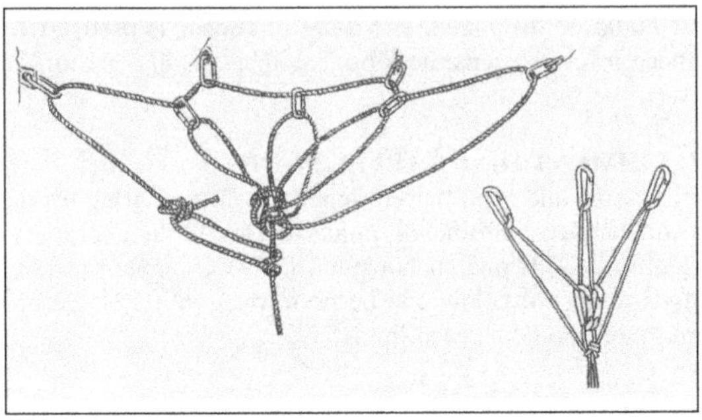

**Figure 2-2. Self-equalizing anchor system.**

(4) In an emergency, double the rope and place it behind or through a secure anchor point, or tie a runner around an anchor point with a snaplink inserted and place the rope through the snaplink. To preclude a rappeller from sliding off the end of the rappel lane, tie a double figure eight (square knot or double fisherman's knot) at the bottom end of the rope with both ends.

b. If a rappel lane is *greater* than half the rope length, the rappeller may apply one of the following techniques:

(1) Using two ropes, tie a round turn anchor bowline around a primary anchor point. Take the remaining rope (the tail from the primary anchor bowline) and tie another round turn anchor bowline to a secondary anchor point. The secondary anchor point should be in a direct line behind the primary anchor point. The anchor can be either natural or artificial. The ends of the rappel lane ropes should be offset by 15 centimeters (6 inches) so that the rope ends feed freely through the rappeller's snaplink.

(2) Using two ropes, establish a three-piton anchor system using a bowline-on-a-bight knot (or figure-eight-on-a-bight knot) and tie off on the long-standing end with a round turn anchor bowline.

(3) In an emergency, use two ropes and tie the two ends together with a joining knot. Place the joined ropes behind or through an anchor point, or tie a runner around an anchor point with a snaplink inserted and place the joined rope through the snaplink. Offset the joining knot to the left or right of the anchor. Tie off the bottom end of the rope with a joining knot to prevent a rappeller from sliding off the end of the rappel line.

c. Due to the length of the rappel, the rappel rope may not reach the anchor. If the rope is used to tie the knots, it may be too short to accomplish the rappel.

(1) When using a natural anchor—

(a) Tie a sling rope, runner, or another rope around the anchor with a round turn anchor bowline.

(b) Tie a fixed loop (figure eight or butterfly) in one end of the rappel rope, which is attached to the round turn around the anchor through the two snaplinks (opposing gates).

(2) When using an artificial anchor—

(a) Tie off a sling rope, runner, or another rope to form a loop.

(b) Put the loop through the snaplinks that are attached to the artificial anchor point.

(c) Bring the bottom of the loop up and connect it to the snaplinks that are between the artificial anchors (chocks, pitons, or bolts).

(d) Grasp the snaplinks that are between the chocks/pitons and pull them down and together.

(e) Tie a fixed loop (figure eight or butterfly) in the end of the rappel rope and connect this to the snaplinks that have been pulled together.

**Notes:** 1. Rerouted figure-eight knots can be used instead of bowlines.
2. Runners may be used from one or more anchor points.

d. Backfeed (stack) the rope to ensure it does not snarl when thrown. Take off one wrap at a time and let it fall to the ground, ensuring that no kinks, knots, or twists occur that might hinder the rope from feeding out. When the rope is backfed, anchor off one end of the rope. The two methods used to throw the rope are underhand and overhand. Use the overhand method when trees or shrubs are on or near the rappel point.

e. Set up a retrievable rappel point by applying one of the following techniques:

(1) When the rappel is *less* than half the total length of the rope, double the rope. Place the rope around the primary anchor with the bight formed by the midpoint. Join the tails of the rappel rope and throw the rope over the cliff. Tie a clove hitch knot around a snaplink just below the anchor point ensuring the locking bar inside the snaplink points away from the gate opening end and faces uphill. Snap the opposite standing portion into the snaplink. Upon reaching the bottom of the cliff, pull on the part of the rope to which the snaplink is secured to allow the rope to slide around the anchor point.

(2) When the length of the rappel is *greater* than half the length of the rope used, join two ropes around the anchor point (double fisherman's knot or square knot). Adjust the joining knot so that it is away from the anchor. Tie a clove hitch knot around a snaplink just below the anchor point ensuring the locking bar inside the snaplink points away from the gate opening end and faces uphill. Snap the opposite standing portion into the snaplink. Upon completion of the rappel, pull the rope to which the snaplink is secured to allow the rope to slide around the anchor point.

**WARNING**

**When setting up a retrievable rappel using only a primary point, take care in selecting the point. Ensure soldiers have a safety line when approaching the rappel point, with only the rappeller going near the edge.**

## 2-5. TYPES OF RAPPELS

Types of rappels include body rappel, hasty rappel, seat-hip rappel, seat-shoulder rappel, Australian rappel, and buddy-evacuation. (Seat-hip rappel and Australian rappel are described in Chapter 1, paragraphs 1-10 and 1-11.)

a. **Body Rappel.** Face the anchor point and straddle the rope. Pull the rope from behind and run it around either hip, diagonally across the chest, and back over the opposite shoulder (Figure 2-3). Then, run the rope to the brake hand, which is on the same side of the hip that the rope crosses (for example, the right hip to the left shoulder to the right hand). Lead with the brake hand down and face slightly sideways. The foot corresponding to the brake hand precedes the guide hand at all times. Keep the guide hand on the rope above to guide—not to brake. Lean out at a sharp angle to the rock. Keep the legs spread well apart and relatively straight for lateral stability and the back straight to reduce friction. Turn the BDU collar up to prevent rope burns on the neck. Wear gloves, and use other clothing to pad the shoulders and buttocks. To brake, lean back and face directly toward the rock area so the feet are horizontal to the ground.

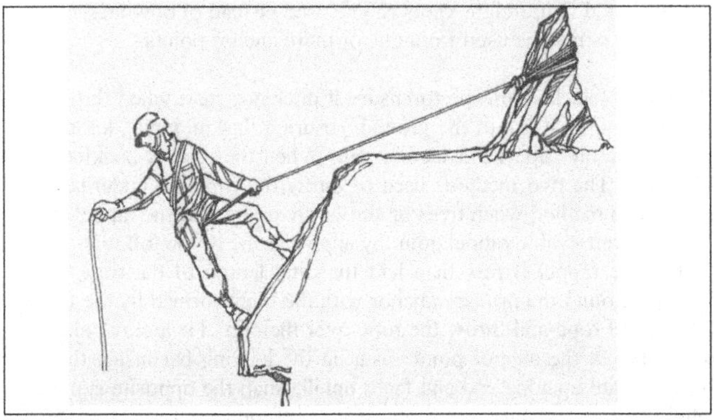

**Figure 2-3. Body rappel.**

b. **Hasty Rappel.** Facing slightly sideways to the anchor, place the ropes horizontally across the back (Figure 2-4). The hand nearest to the anchor is the guide hand, and the other is the brake hand. Wear gloves to prevent rope burns. To stop, bring the brake hand across in front of the body locking the rope. At the same time, turn to face up toward the anchor point. Use this rappel only on moderate rock pitches. The hasty rappel's main advantage is that it is easier and faster than the other methods, especially when the rope is wet.

**Figure 2-4. Hasty rappel.**

**Figure 2-5. Seat-shoulder rappel.**

c. **Seat-Shoulder Rappel.** To hook up for the seat-shoulder method, face the rappel point (Figure 2-5). Snap into the rope that passes up through the snaplink. Bring the rope over one shoulder and back to the opposite hand (left shoulder to right hand). Use the same technique in the descent as in the body rappel. This method is faster than the body rappel, less frictional, and more efficient rappellers with packs and during night operations.

d. **Buddy-Evacuation Rappel.** Use the buddy-evacuation rappel to evacuate an injured soldier from a cliff or steep terrain. Face the cliff and assume a modified L-shape body position to compensate for the weight of the victim on the back. The victim is top-rope belayed from above, which provides the victim with a point of attachment to a secured rope.

**Note:** To use this rappel, the victim must be conscious.

(1) The method for securing a victim to a rappeller's back is described below:

(a) To secure the victim to the carrier's back with a rope, the carrier ties a standard rappel seat (brake of choice, depending on the injury) and rests his hands on his knees while the victim straddles his back (Figure 2-6.)

(b) A 4.2-meter (14-foot) sling rope is used. A 45-cm (18-inch) tail of the sling is placed on the victim's left hip. (This method describes the procedure for a seat-hip rappel with right-hand brake.)

(c) The long remaining end of the sling rope is routed under the victim's buttocks and passed over the victim's and carrier's right hip. The rope is run diagonally, from right to left, across the carrier's chest, over his left shoulder, and back under the victim's left armpit.

(d) The rope is then run horizontally, from left to right, across the victim's back. The rope is passed under the victim's right armpit and over the carrier's right shoulder.

**Figure 2-6. Patient secured to carrier's back.**

(e) The rope is run diagonally, from right to left, across the carrier's chest and back across the carrier's and victim's left hip.

(f) The two rope ends should now meet. The two ends are tied together with a square knot and overhand knots.

(g) The knot is positioned on the victim's left hip. The carrier's shoulders should be padded to prevent cutting by the rope.

(2) An alternate method is to use two pistol belts hooked together and draped over the carrier's shoulders. The victim straddles the carrier, and the belay man secures the loose ends of the pistol belts under the victim's buttocks. Slack in the pistol belt's sling should be avoided, since the carrier is most comfortable when the victim rests high on his back. (See FM 8-35.)

(3) A large rucksack can be slit on the sides near the bottom so that the victim can step into it. The victim is belayed from the top with the carrier conducting a standard rappel. The carrier wears the rucksack with the victim inside.

**Note:** Rucksacks are not designed to support the weight of a soldier and his gear. Therefore, this technique is used only as a last resort.

(4) A casualty secured to a carrier, as described above, can be rappelled down a steep cliff using a seat-shoulder or seat-hip rappel. The casualty's and rappeller's shoulders should be padded where the sling rope and rappel lines cross if a seat-shoulder rappel is used. The buddy team should be belayed from above with a bowline tied around the victim's chest under his armpits. The belay rope must run over the rappeller's guide hand shoulder (Figure 2-7).

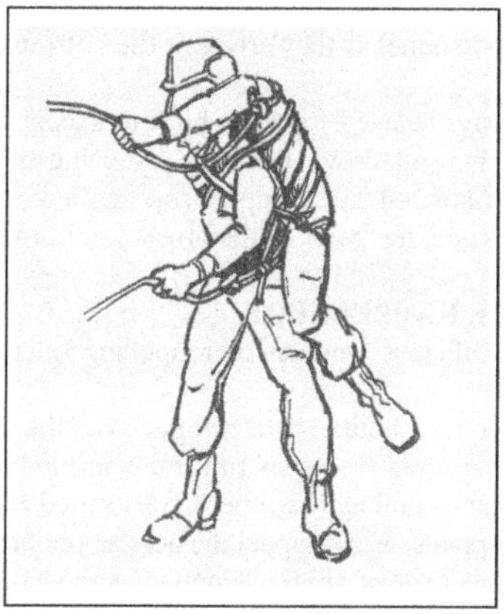

**Figure 2-7. One-man carry.**

## 2-6. RAPPELLING PROCEDURES

Proper rappelling procedures must be followed to ensure the safety of personnel.

a. The rappel lane NCO is tied off to a separate anchor point for safety, if possible. He—

(1) Ensures that the anchors are sound and the knots are properly tied.

(2) Ensures that loose rock and debris are cleared from the loading platform.

(3) Allows only one man on the loading platform at a time and ensures that the rappel point is orderly run.

(4) Ensures that each man is properly prepared for the particular rapel: gloves on sleeves down, helmet with chin strap fastened, gear prepared properly, and rappe seat and knots correct (if required). He also ensures that the rappeller is correctly hooked up to the rope and is aware of the proper braking position.

(5) Ensures that the proper signals or commands are used.

(6) Dispatches each man down the rope.

(7) Is the last man down the rope.

(8) Ensures that the rope does not run over sharp edges, and pads the rappelling surfaces as necessary to protect the rope.

(9) Ensures that the rope reaches the bottom or is at a place where additional rappels can be made.

b. The first rappeller down—

(1) Selects a smooth route for the rope that is clear of sharp rocks.

(2) Conducts a self-belay with a prusik sling tied from himself to the rappel rope.

(3) Clears the route and places loose rocks, which the rope may dislodge, far enough back on ledges to be out of the way.

(4) Ensures the rope reaches the bottom or is at a place from which additional rappels can be made.

(5) Ensures that the rope runs freely around the rappel point when pulled from below.

(6) Clears the rappel lane by straightening all twists and tangles from the ropes.

(7) Belays subsequent rappellers down the rope.

(8) Takes charge of personnel as they arrive at the bottom (off-loading platform).

**Note:** A rappeller is always belayed from the bottom except for the first man down. The first man belays himself down the rope by using a safety line attached to his rappel seat that is hooked to the rappel rope with a prusik knot. As the first man rappels down the rope, he "walks" the prusik knot down with him.

## 2-7. DUTIES OF THE RAPPELLER

To ensure the safe and efficient conduct of rappelling operations, each rappeller must know the duties of his job.

a. Each rappeller down shouts, "Off rappel" (if the tactical situation permits), untangles the ropes, and ensures the ropes run freely around their anchors. When silence is needed, a planned signal of pulling the rope is substituted for the voice signal. After the rope is cleared and the rappeller is off rappel, he acts as the belayer for the next rappeller.

b. All rappellers inspect ropes often when many soldiers are rappelling.

c. The last rappeller to descend constructs a retrievable rappel point and rappels down. Then he gently pulls the rope to prevent the rising rope end from entangling with the other rope. He stands clear of the falling rope and any rocks that it may dislodge.

**Note:** Soldiers wear gloves for all types of rappels for protection from rope burns. Also, bounding rappels are discouraged since this stresses the anchor and causes undue wear and friction on the rope.

d. Rappellers descend in a smooth, controlled manner. The body forms an L-shape with the feet shoulder-width apart, legs straight, and buttocks parallel to the ground. When carrying equipment or additional weight, a modified L-shape is used with the legs slightly lower than the buttocks to compensate for the added weight. The rappeller's back is straight. He looks over the brake shoulder. The guide hand is extended on the rope with the elbow extended and locked. The rope slides freely through the guide hand, which is used to adjust equipment and to assist balance during the descent. The rappeller grasps the rope firmly with the brake hand and places it in the small of his back. Releasing tension on the rope and moving the brake hand out to his rear at a 45-degree angle regulates the rate of descent. The rappeller never lets go of the rope with his brake hand until the rappel is completed.

e. During training, the lane number must be understood. In combat, a series of tugs on the rope may be substituted for the oral commands to maintain noise discipline. The number of tugs used to indicate each of the commands is IAW the unit SOP.

## 2-8. BELAYER

The belayer assumes a position at the base of the lane about one pace away from the rock area. He ensures that the rappel ropes are even with the ground during rock and tower rappels. The belayer loosely holds the rappel ropes with both hands so as not to interfere with the rappeller but still able to stop the rappeller should he fall. If the rappeller shouts "Falling" or loses control of his brake hand or descent, the belayer immediately stops the rappeller by pulling downward on the rappel ropes. To ensure a firm grip on the rappelling rope, the belayer does not wear gloves. (Because no friction exists between the belayer's hands and the rappelling rope, gloves are not required for safety. However, whether or not to wear gloves is decided by the OIC/NCO.) The belayer watches the rappeller at all times and maintains constant voice or visual contact with the rappeller. The belayer wears a helmet to prevent injuries from falling debris. All commands are spoken loudly and clearly.

CHAPTER 3

# HELICOPTER RAPPELLING

*Helicopter rappelling can provide a means of quick insertion with or without an LZ.*

## Section I. PERSONNEL

The personnel required for helicopter rappelling are the rappel master, the rappel safety officer, the pilot in command, the rappelers, and the belayers.

### 3-1. RAPPEL MASTER

A qualified rappel master is aboard each aircraft. Safety is the rappel masters number one priority. The rappel master—

- Ensures internal communication between the pilot and rappel master, and external communication between the aircraft and the ground.
- Inspects all equipment and uses only authorized, serviceable equipment.
- Inspects and tests all anchor points and knots before the mission starts.
- Ensures that all rappellers receive a safety briefing and the pilots and aircrew receive an air mission brief.
- Ensures that the rappelers are rappel qualified before conducting helicopter rappelling to include tactical rappelling.
- Maintains communications with the pilot at all times.

### 3-2. RAPPEL SAFETY OFFICER

The RSO is a SFC or above who is either air assault or ranger qualified. He has overall responsibility for the safety of all rappellers and ensures that all safety precautions are followed. He maintains communications at all times with the pilot and rappel master through FM radio. He alerts the rappel master and pilot of any unsafe acts.

### 3-3. PILOT IN COMMAND

The pilot in command (PIC) of the aircraft has the following responsibilities:

a. Ensures that the aircrew and all non-aircrew personnel are briefed and understand their responsibilities during rappelling operations, including aircraft safety and action in the event of an emergency.

b. Ensures that the donut ring anchoring device assembly and or aircraft anchor points have been inspected for completeness and functionality, and that they are installed properly.

c. Emphasizes procedural techniques for clearing, recovery, and or jettison of ropes.

d. Keeps the aircraft centered over the target with corrections from the rappel master as required.

### 3-4. RAPPELLER

In addition to the tower qualification requirements outlined in Chapter 1, the individual rappeller must complete advance training under the supervision of a qualified rappel master in order to participate in tactical helicopter rappelling operations. He must—

- Satisfactorily complete three rappels from a helicopter from a height of 60 feet (two rappels with combat equipment and weapon).
- Demonstrate confidence and proficiency in the techniques, procedures, and equipment used in rappelling from a helicopter.
- Know the rappelling equipment used in helicopter operations and any special equipment required for helicopter rappelling.

### 3-5. BELAYER

A belayer is assigned to each rope. He is responsible for walking the rope beneath the helicopter during the descent. (Walking the rope is defined as removing the slack from underneath the helicopter by walking backwards with the rope as the helicopter descends to land.) The belayer makes sure the ropes are not caught on the aircraft skids or tires to ensure a safe landing.

## Section II. TRAINING

Training for helicopter rappelling includes sustainment training and refresher training.

### 3-6. SUSTAINMENT TRAINING

Before conducting helicopter rappel training, the unit or element conducts sustainment training.

a. Review the construction of a rappel seat, equipment to be used, and hook-up procedures.

b. Conduct two rappels on the 34-foot or higher tower wall: one without equipment, one with equipment. Conduct two rappels from the open side of a 34-foot or higher rappel tower.

### 3-7. REFRESHER TRAINING

Refresher training is routinely conducted to maintain acquired skills. Soldiers who have not performed a helicopter rappel during the past six months will undergo refresher training consisting of three satisfactory rappels from a tower (one with weapon and equipment and one executing a lock in) before executing a helicopter rappel.

## Section III. PREOPERATIONAL BRIEFINGS AND SAFETY PROCEDURES

This section discusses safety, medical and communications requirements, and the procedures to follow during unusual conditions (adverse weather/terrain conditions, night operations). Personnel must use sound judgement to determine what action to take depending on the nature and severity of the conditions.

### 3-8. MEDICAL COVERAGE

A qualified and equipped medic will be present to respond to any mishap. Medical transportation must also be available. Absence of a medic, medical equipment, or transportation will terminate training. If the situation warrants and the installation cannot support a MEDEVAC mission, the rappel aircraft may be used as a last resort MEDEVAC vehicle.

## 3-9. COMMUNICATION REQUIREMENTS

During helicopter rappel training, the RSO will have radio communications with the aircraft. Voice communications are required before starting aircraft rappelling. Additionally, the RSO will inform the PIC to stop operations if an unsafe condition develops.

## 3-10. ADVERSE WEATHER/TERRAIN CONDITIONS

Rappel operations will not be conducted under the following conditions

- Ambient temperature is 30 degrees Fahrenheit or less.
- Winds in excess of 30 knots.
- Lightning strikes within 1 nautical mile of rappelling operations.
- Wind chill factors caused by the helicopter's rotor wash or extraction cruise air speeds, which could cause cold weather injuries.
- Water or ice on the rope inhibiting the ability of the rappellers to control their descent.
- The rope is exposed to the elements for a sufficient length of time to freeze, thereby reducing its tensile strength.
- Blowing particles produced by rotor wash causes the aircrew or the rappel master to lose visual contact with the ground.

## 3-11. NIGHT OPERATION REQUIREMENTS

The following requirements are necessary for night rappelling operations.

a. One chemlight will be attached to the end of the rope and one on each rappeller.

b. One chemlight will be secured to the attachment point of the rope.

c. Night vision goggles (NVG) will not be worn by rappellers during the descent. Aircrew members will wear NVG as required during night operations. NVG lighting criteria will be IAW Army regulations, specific aircraft aircrew training manuals, unit SOPs, or the tactical environment.

## 3-12. SAFETY BRIEFING

The following safety measures are enforced.

a. Loose clothing and equipment are secured.

b. Helmets are worn with chin straps fastened.

c. Rappellers wear identification tags and earplugs, carry identification cards, and role down their sleeves.

d. Weapons are slung diagonally across the back with the muzzle pointing down on the guide hand side.

e. All seats and rappelling equipment must be inspected by a rappel master before rappelling.

f. No running is allowed within 50 feet of the aircraft.

g. Personnel approach and depart the helicopter from the front and forward of the rear of the cargo doors. When approaching or departing the helicopter, personnel bend their bodies forward at the waist to ensure clearance of the rotor blades. At no time will personnel go near the rear of the aircraft.

h. Upon boarding the aircraft, the rappeller sits or kneels down, hooks up, and applies his brake hand to the small of his back.

i. While in the helicopter, the rappeller maintains eye-to-eye contact with the rappel master and receives commands from him.

j. The rappeller ensures that he has a belayer on his rope at all times when conducting training at a hover site.

k. During the descent, the rappeller maintains eye-to-ground contact.

l. If the rappeller sees his rope coming off the ground or sees that his belayer has lost control of his rope, he immediately brakes and executes a lock-in. He then waits for commands from the rappel master.

m. The rappeller brakes once every 30 feet during descent.

n. The belayer does not wear gloves. He keeps both hands on the rope and his eyes on the rappeller at all times.

o. All rappelling will be conducted using a double strand of rope.

## Section IV. DEPLOYMENT OF ROPES

Deployment of the ropes from a helicopter is a critical task. It can cause a planned rappelling operation to fail, or it can increase the time required to conduct the operation. This is due to the likelihood of the ropes becoming entangled. To prevent this, ropes must be deployed using a positive control technique. Two of the techniques that may be used are the *deployment bag* technique and the *log coil* technique.

### 3-13. DEPLOYMENT BAG TECHNIQUE

The rappeller places the deployment bag (D-bag) (standard) on a flat surface with the stow loop facing upward. If the D-bag still has a static line, the rappeller removes it by cutting the static line where it attaches to the bag. A 12-foot section of the static line can be used as a safety line in the helicopter.

a. Use a sandbag (about one E-tool spade full of sand) as a weight in the deployment system. Roll the sandbag into a rectangular shape, tie it, and then place it in a small plastic bag, rolled and taped. Place a retaining band over the middle of the weight, and place the weight on top of the D-bag. The weight should be about the same width as the D-bag.

b. Lay out and inspect the two ropes. The working ends (closest to the D-bag) should be even. Place a round turn with the two working ends of the ropes on the weight's retaining band. Working on top of the deployment bag, start forming figure eights. The stack should consist of 8 to 10 figure eights, one on top of another—do not exceed the width of the D-bag. Then, starting from either side, center a retaining band over the stack. Ensure that it is over all the figure eights in the stack. Repeat the process each time, placing one stack in front of the other. Continue until about 10 feet of rope remains. If one rope is shorter than the other, the end of the shorter rope should be about 10 feet from the last stack.

c. For a primary anchor point, measure down about 4 feet from the end of the shortest rope. Using both ropes, tie a bowline without a half hitch. This knot is the primary anchor point.

d. For a secondary anchor point, tie an end-of-line bowline with a half hitch toward the end of the ropes. The dressed knot should be 18 to 22 inches from the primary anchor point knot when using the UH-1H helicopter and 22 to 30 inches when using the UH-60 helicopter. If excess rope remains, S-fold and tape it between the two knots. Ensure that 3 to 6 feet of rope remain from the last stack of figure eights.

e. To place the ropes in the D-bag, remove the bag from under the stack of rope. Place the weight into the bottom of the D-bag, and place all the stacks of figure eights, in the order they were made, one on top of the other into the bag.

f. Two left, two right, and two rear loops are in the opening of the D-bag. ***Do not use the two rear loops***. Close the flap on the D-bag and push the loops through the aligning holes on the flap. Using an 8-inch piece of gutted 550-pound cord, tie the two loops together with a square knot and two hitches. Ensure that two 120-foot ropes are coming out of the center of the flap—not to one side. Repeat the process and tie the right loops together. Wrap the excess rope lengthwise around the bag.

**Note:** The sizes of the loops of the bowline should be no larger than an average size fist.

### 3-14. LOG COIL TECHNIQUE

The rappeller lays the running end of the double rope along the length of the coiling log (Figure 3-1). He then coils the double rope around both the running end of the rope and the coiling log. The rope must be coiled evenly and tightly.

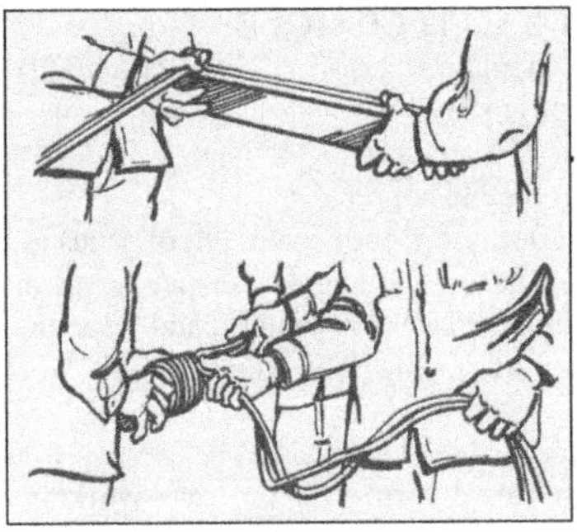

**Figure 3-1. Coiling Log**

## Section V. RAPPELLING OPERATIONS FOR UH-1H IROQUOIS HELICOPTER

The UH-1H helicopter provides a safe, stable aerial platform from which to conduct rappelling operations when landing is not feasible.

### 3-15. CHARACTERISTICS

The UH-1H is a single-engine, medium-speed, single-main rotor helicopter that can transport eight rappellers, one rappel master, and one three-man crew. It has a lift capacity of 2,300 pounds and is equipped with several floor-mounted tie-down fittings, seven of which are used during rappelling operations.

## 3-16. RIGGING OF UH-1H HELICOPTER FOR RAPPELLING

The rappel master performs the following to rig the UH-1H helicopter.

a. Removes all seats.

b. Locks the doors in the open position. If no locks are present, removes doors. Also, he removes small cargo doors.

c. Pads and tapes all sharp edges on the floor, door ledge, and all protrusions on the skids. Ensures each door ledge has a scuff pad to protect the rope from contacting the metal door ledge.

d. Secures the donut ring to the center of the floor of the helicopter. The donut ring has six snaphooks numbered clockwise, with 12 o'clock being toward the front of the helicopter. Positions the clamps of the donut ring toward the aft end of the helicopter. Ensures the front two snaplinks and rear two snaplinks are facing outside of the donut ring and hooks them to the tie-down ring with the snaphook facing down. Ensures the center two snaphooks face into the donut ring and are connected to the floor tie-down ring with the gate facing down (Figure 3-2). Attaches the free-floating safety ring to the center floor tie-down ring using a seventh snaphook.

**Note:** Using this technique has caused equipment failure in many aircraft. The cable clamp nuts must be inspected and certified as airworthy by the pilot or maintenance crew before flying the mission.

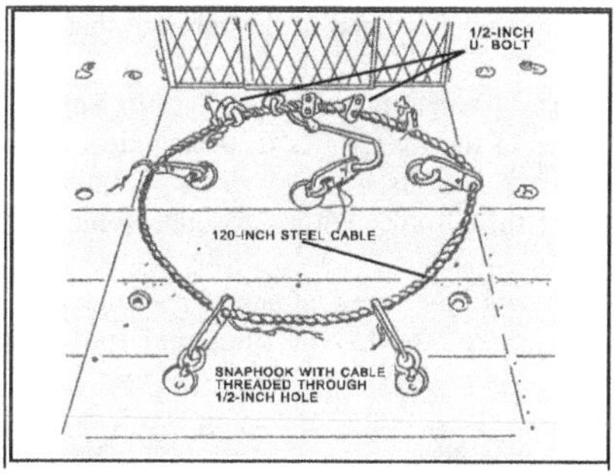


**Figure 3-2. Donut ring attached to the floor of the helicopter.**

## 3-17. CONSTRUCTION OF ANCHOR POINTS

In addition to the usual equipment requirements for rappelling operations, an anchor assembly is fabricated. This anchor assembly is commonly known as a *donut ring*. The secondary anchor point is a *floating safety ring*.

a. **Donut Ring.** This primary (or No. 1) anchor point for the rappelling ropes is constructed from a 1/2-inch steel cable with a steel wire core. The cable is 120 inches long, consists of 6 strands (18 wires per strand), and has a tensile strength of 21,000 pounds (Figure 3-2). The completed donut ring has a tensile strength of 3,000 pounds.

(Figure 3-3 shows the donut ring and rappel rope connection in a helicopter.)

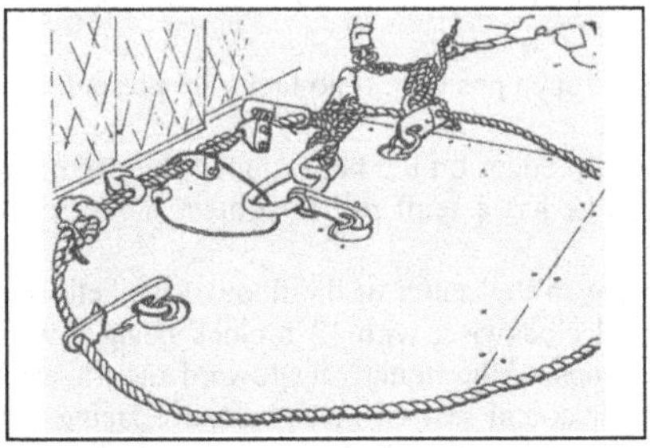

**Figure 3-3. Donut ring and rappel rope connection in a helicopter.**

(1) Thread six parachute static line snaphooks onto the 120-inch steel cable so that four snaphooks are facing out with the gates down, and the center two snaphooks are facing in with the gates down. Drill the end portion of each snaphook to make a 5/8-inch diameter hole, and thread the cable through the holes in the snaphooks.

(2) Overlap the ends of the 120-inch cable 20 inches to form a circle. Secure the ends with four 1/2-inch U-bolts placed at 2- to 3-inch intervals.

(3) Attach two U-bolts to each dead end of the cable so that the bolts engage the dead end.

(a) Before torquing the U-bolts, position a 12-inch length of chain or 1/8-inch diameter cable on the center of the overlapped 120-inch steel cable so that it remains in position between the two center U-bolts.

(b) Tighten each nut of the U-bolts with a torque wrench (if possible) to 40-foot pounds (480-inch pounds).

(c) After the U-bolt clamps have been attached and tightened, fasten a steel plate (drill to fit) over the open end of the U-bolt studs and spot weld in place to prevent loosening.

b. **Floating Safety Ring**. The floating safety ring is referred to as the secondary (or No. 2) anchor point for the rappelling ropes (Figure 3-4). The snaplink at the end of the rappelling rope is hooked to this connection. Either of the following two types of floating safety rings may be used:

(1) *First method.* Thread an elliptical rappelling ring through the free end of the keeper chain (cable). It is constructed of cold-rolled steel that is 1/4-inch in diameter, with inside dimensions of 2 1/4 inches (minor axis) and 4 inches (major axis). Thread a seventh parachute static line snaphook onto the ring before welding. Weld this ring together so that it withstands a force up to 3,000 pounds.

(2) *Second method.* Attach two snaplinks to the aircraft tie-down ring in the center of the donut ring. Insert the first snaplink through the free end of the keeper chain (cable) and the tie-down ring with the gate down. Insert the second snaplink through the free end of the keeper chain (cable) and the tie-down ring with the gate up. Tape the snaplink

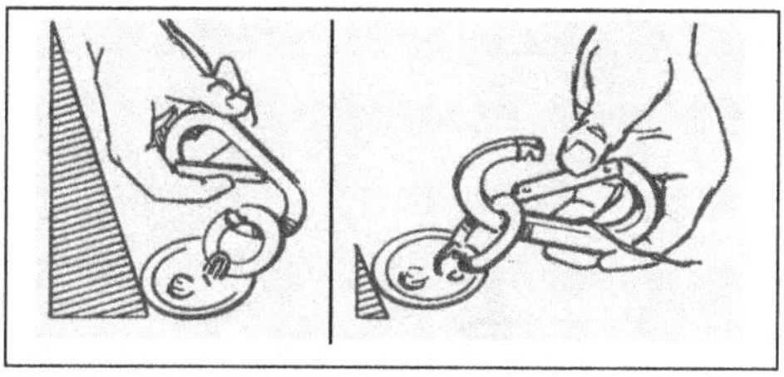

gates closed with masking tape. Then, tape the snaplinks together to ensure that the snaplink gates are on opposite sides of each other.

**Figure 3-4. Floating safety ring formed with two snaplinks.**

c. **Rappel Rope Anchor Points.** The rappelling rope is connected to the floating safety ring and the donut ring as follows:

(1) *Number 1 anchor point (donut ring).* Using a bight about 5 feet from the end of the standing part of the rope, make one turn through the snaplink forming a round turn. The bowline is the preferred method of attaching rappel ropes to snaplinks or anchor systems. The round turn with two half hitches is a reliable means of attaching ropes to anchor systems. Secure the round turn to the snaplink with two half hitches. Make the connection to the donut ring by attaching the snaplink gate upward. Ensure the gate faces upward with the opening away from the knot.

(2) *Number 2 anchor point (floating ring).* Attach the No. 2 snaplink the same as the first with some exceptions. Using a bight about 2 feet from the end of the standing part of the rope, connect the snaplink to the rope the same as the first connection. Tape the end of the standing part of the rope and the knots with masking tape (or green engineer tape) to secure them in place. Make the connection to the floating safety ring the same as the connection to the donut ring (Figure 3-5). Four ropes can be connected to the floating safety ring using two snaplinks (Figure 3-6).

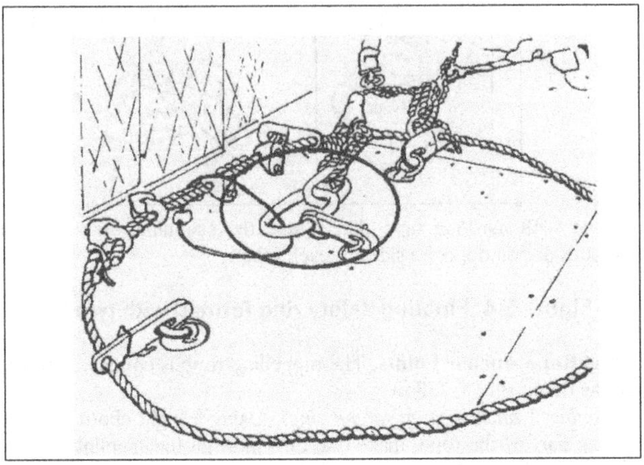

**Figure 3-5. Rappel rope connection using two snaplinks for the floating safety ring.**

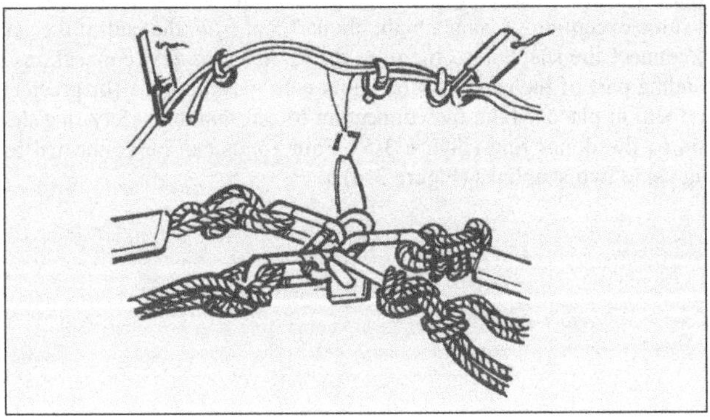

**Figure 3-6. Four rappel ropes connected to the floating safety ring (two snaplinks).**

## 3-18. SEATING ARRANGEMENTS AND LOADING TECHNIQUES

Rappelling operations are executed quickly and safely to ensure the success of the operation in a combat or training environment. All rappellers must know the seating arrangements and the procedures for loading (Figure 3-7). They must be rehearsed under the supervision of the rappel master before conducting the operation.

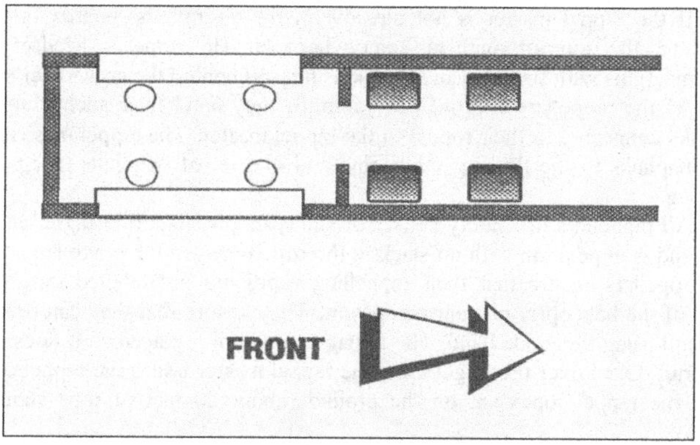


**Figure 3-7. Seating arrangement.**

a. The first four rappellers are hooked up to the donut ring and use safety belts while in flight. The remaining rappellers use seat belts or safety straps. For safety emphasis during flight, each rappeller's brake hand is in position with no slack in the rope between the brake hand and the donut ring. The rappeller ensures that his rappelling rope is not tangled on any part of the interior of the helicopter or his equipment. He ensures that the coiling log or D-bag is located in such a position that he drops the rope with ease, using his guide hand. The coiling log or D-bag is secured with half of it under the rappeller's inside leg and the other half on top of his outside leg.

b. The rappellers should board the aircraft and sit along the leading edge of each door. They hook up with the rappel seat square knot on the inboard side of the aircraft and the brake hand on the outboard side. The rappel master sits in the center to aid visual inspection of personnel and equipment.

c. For a flight time of less than five minutes and for training and operations where speed is a factor, the rappel master directs the rappellers to be seated in the door facing outward with their feet on the skid. In a tactical situation and before the helicopter comes to a hover, the rappellers assume an upright position facing inside the helicopter.

d. The aircraft is cleared as fast as possible. This reduces the time that the aircraft is in the high-hover attitude to a minimum.

## 3-19. RAPPELLING PROCEDURES

The rappel master ensures all personnel follow the rappelling procedures for the safe and efficient execution of the operation. These procedures are rehearsed under the supervision of the rappel master.

a. Rappellers (eight maximum) are identified, and rappel masters tell them which door of the aircraft to exit. Rappellers approach the aircraft once it is firmly on the ground, and the crew chief signals when it is safe to approach. The approach to the aircraft should be about 45 degrees to the nose of the aircraft allowing the pilot to see the rappeller.

b. If the rappel master is not already on the aircraft, he boards first and secures himself to the aircraft with a safety harness. He dons a headset to maintain communications with the pilot and positions himself behind the crew chief's chair.

c. As the rappellers approach the aircraft, they hand their anchor attaching points (snaplinks connected to their ropes) to the rappel master. The rappel master connects one set of snaplinks to the floating safety ring and one set of snaplinks (the primary) to the donut ring.

d. All rappellers use safety belts. For safety emphasis during flight, each rappeller's brake hand is in position with no slack in the rope between the brake hand and the donut ring. Rappellers ensure that their rappelling ropes are not tangled on any part of the interior of the helicopter or their equipment. They ensure that they can drop their D-bag or log coil using the guide hand. The D-bag or log coil is placed so it does not fall out of the aircraft. Once over the target area, the rappel master issues the rappel commands and ensures the rappel ropes are on the ground (about 20 feet of rope should be on the ground).

e. On the command GO, the rappeller flexes at the knees and vigorously pushes away from the skid gear, allowing the rope to pass through his brake hand and guide hand. The descent is accomplished smoothly at a rate of about 8 feet per second, avoiding jerky stops. The rappeller initiates his braking action slowly when he is about halfway to the ground. The safety officer or NCO ensures that at least a five-second delay is maintained between each rappeller. Another technique used to deploy rappellers is to issue the command GO to the rappellers diagonally opposite each other. In the early phase of performance, the rappeller's feet are together (particularly if over woods or jungle) and his guide hand is used (not the brake hand) to remove any entanglements.

f. Upon reaching the ground, the rappeller clears the rappel rope through the snaplink (or rappel ring) until the rope is free. If other rappellers are to follow on the same rope, the rappeller on the ground separates or untwists the ropes and becomes the belayer for each subsequent rappeller. If ropes are to be released from the aircraft following the last rappeller, untwisting and separating the ropes is not necessary. A two-second interval is maintained between exit groups (two men exiting at the same time).

g. The rappel master releases the rope from the donut ring after he confirms (by visual inspection) that the rappeller is off rappel. He then drops the rope away from the helicopter.

## 3-20. RAPPELLING COMMANDS

Helicopter rappelling is conducted in a noise-filled environment. Each rappeller is trained to know the rappel commands and to understand the actions required of him during the execution of each command. Due to the noise created by the helicopter, all rappel commands should be accompanied by arm-and-hand signals. All commands are issued by the rappel master and confirmed orally by the rappellers.

a. The helicopter rappelling commands are as follows:

(1) GET READY. The rappeller checks his combat equipment, looks toward the donut ring, and pulls the rope to check the anchor point connection. He then checks his

snaplink to ensure the rope is properly seated in the snaplink on his rappel seat.

(2) THROW ROPE. The rappeller drops his deployment bag out and away from the helicopter with his guide hand. He ensures that the rope does *not* fall between the side of the helicopter and the skid. The rappel master ensures the rope is touching the ground and is free of tangles and knots. The rappeller then replies, "Rope okay."

(3) SIT IN THE DOOR. The rappeller swings his legs to the outside of the helicopter and takes up a sitting position. (This command applies only to the UH-1H helicopter.)

(4) POSITION. The rappeller pivots 180 degrees on the skid. His feet are shoulder-width apart, the balls of his feet are on the skid, his knees are locked, his body is bent forward in an L-shape position, and his brake hand is on his buttock. The rappel master makes a visual inspection of the snaplink and rappel ring.

(5) GO. The rappel master points at the rappeller to exit. One rappeller from each side rappels at the same time. The rappel master gives the command GO to the rappellers that are diagonal to each other—for example, the front left rappeller and the rear right rappeller. The rappeller flexes his knees and jumps backward, letting the rope run through the brake hand and guide hand.

b. The arm-and-hand signals that are used with rappelling commands are as follows:

(1) GET READY (Figure 3-8). The rappel master extends both arms to the front with fists clenched and thumbs pointing upward.

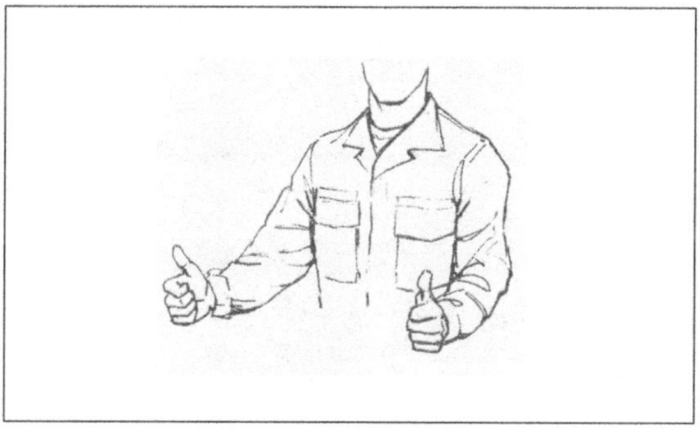

**Figure 3-8. Arm-and-hand signal for GET READY.**

(2) SIT IN THE DOOR (Figure 3-9). The rappel master extends both arms to the front with elbows locked, fingers extended, thumbs running along the index fingers, and palms facing downward. He bends slightly at the waist so that the arms are below the waist. Then, he moves both arms in a crisscross waving motion, alternating left over right and right over left.

**Figure 3-9. Arm-and-hand signal for SIT IN THE DOOR.**

(3) THROW ROPES (Figure 3-10). The rappel master extends both arms to the front with elbows locked, fist clenched and index fingers extended. He points at respective rappellers, brings the forearms to an upright position and then down to the elbow-locked position.

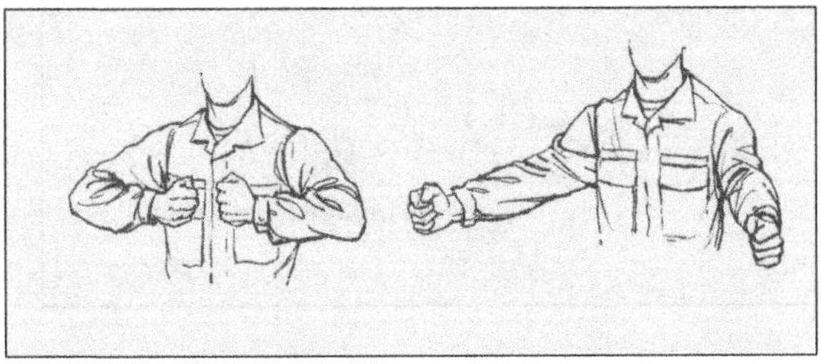

**Figure 3-10. Arm-and-hand signal for THROW ROPES.**

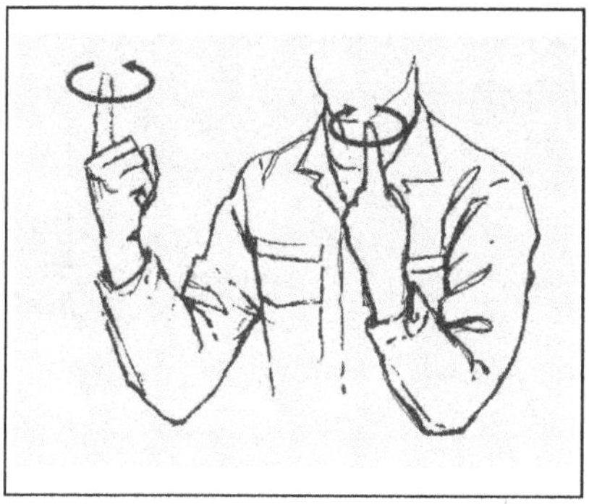

**Figure 3-11. Primary arm-and-hand signal for POSITION.**

(4) POSITION. The two arm-and-hand signals for POSITION are:

(a) Primary signal (Figure 3-11). The rappel master extends both arms to the front, elbows bent, forearms pointed upward, fists clinched, and index fingers pointing upward. He makes a circular motion with both forearms rotating in opposite directions.

(b) Alternate signal (Figure 3-12). The rappel master extends both arms to the front with elbows locked, fists clenched, and index fingers extended. He bends at the waist so that his arms are below his waist and makes a circular motion with arms rotating in opposite directions.

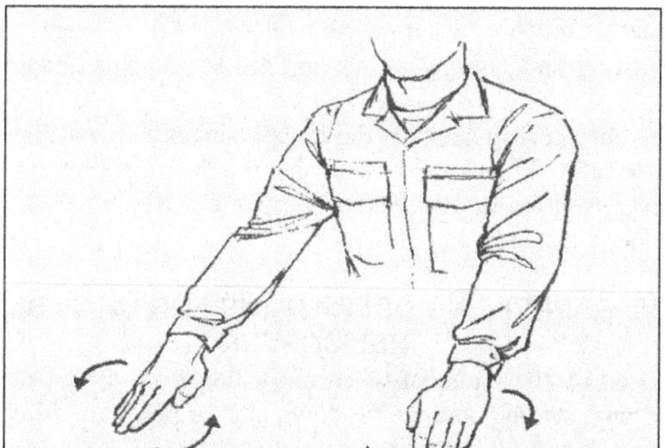

**Figure 3-12. Alternate arm-and-hand signal for POSITION.**

(5) GO (Figure 3-13). The rappel master extends an arm with elbow locked, fingers extended, and thumb along the index finger, and he points directly at the rappeller.

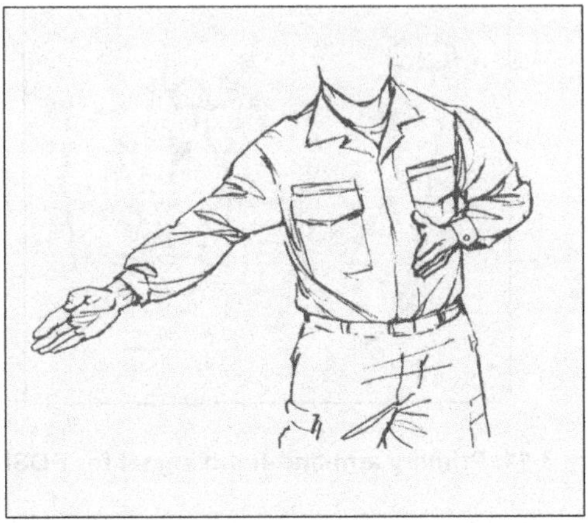

**Figure 3-13. Arm-and-hand signal for GO.**

## 3-21. INSPECTION AND SAFETY CONSIDERATIONS

The rappel master and pilot (or his representative) conduct a joint inspection of the aircraft to ensure the safety of all personnel and serviceability of equipment.

a. Cargo doors are locked in the open position or cleared for closing, depending on the mission.

b. All loose objects in the cargo compartment are removed or secured forward.

c. Sharp edges or protrusions on the cargo floor and door ledges are taped.

d. The donut ring and floating safety ring are serviceable and properly attached to tie-down fittings.

e. A headset and intercom jack for the rappel master are available and operational.

f. Serviceable safety harnesses are available for the rappel master and crew chief.

g. The recovery rope is installed and properly stowed.

h. Unused floor rings are taped down.

# Section VI. RAPPELLING OPERATIONS FOR UH-60 BLACKHAWK HELICOPTER

The techniques used by the rappeller when rappelling from different aircraft are similar. However, positioning, seating, and the tie-down anchor point are different. Each rappeller is well trained on each aircraft before conducting a rappel.

## 3-22. CHARACTERISTICS

The UH-60 is a twin-engine, medium-speed, single-main rotor helicopter that can transport 10 rappellers, 1 rappel master, and 1 four-man crew when the center row of troop seats is removed. It is equipped with four 4,000-pound (load limit) cabin ceiling tie-down fittings or rings that are located in the ceiling of the troop/cargo compartment. These fittings are used as the primary rappelling rope anchoring points. The UH-60 is also equipped with eight 3,500-pound (load limit) cargo restraint net rings. Four of the eight rings are located in the ceiling of the troop/cargo compartment. These rings are used as the secondary rappelling rope anchoring points.

## 3-23. RIGGING OF UH-60 FOR RAPPELLING

The rappel master rigs the UH-60 helicopter by performing the following:

a. Locks both cargo doors in the open position.

**Note:** For arctic or other cold-weather operations, or during flights of long duration, the cargo doors are closed and locked until the time specified for opening them.

b. Removes the center row of troop seats.

c. Tapes any sharp edges or protrusions on the cargo floor and door ledges that may come in contact with the rappeller or the rappelling rope.

d. Stows loose equipment forward in the cargo compartment.

e. Extends the rappel master's intercom cord to the rear over the aft utility drain line and tapes the cord to the overhead troop seat support tube.

f. Installs the floor restraint provisions for rappellers No. 1 through 6.

g. Rigs and connects rappelling ropes to the aircraft's primary and secondary anchoring points.

(1) *Primary anchor points.* Ties a bowline with a half hitch about 4 feet from the standing end of the rope. Attaches the two primary snaplinks to the respective cabin tie-down fitting ring with gates facing in the opposite directions (Figures 3-14 and 3-16)

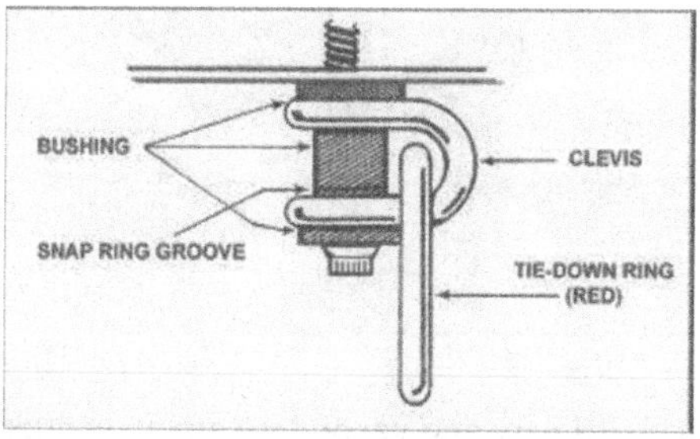


**Figure 3-14. Primary snaplink attaching point.**

(2) *Secondary anchor points.* Ties a bowline with a half hitch about 1 1/2 feet from the standing end of the rope. Attaches the secondary snaplink to the adjacent overhead cargo restraint net ring (Figures 3-15 and 3-16).

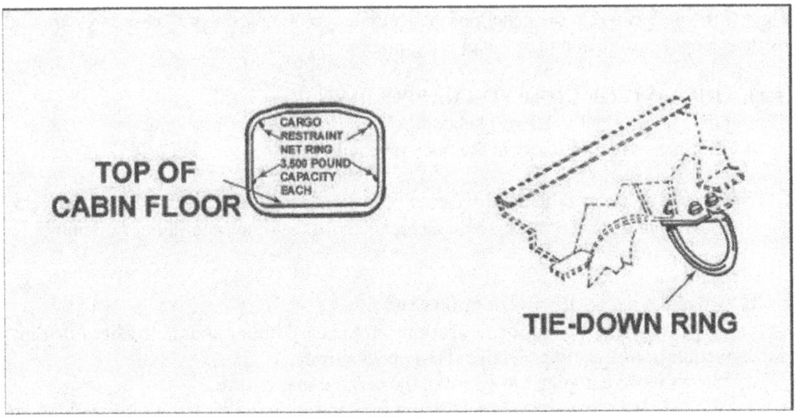


**Figure 3-15. Secondary snaplink attaching point.**

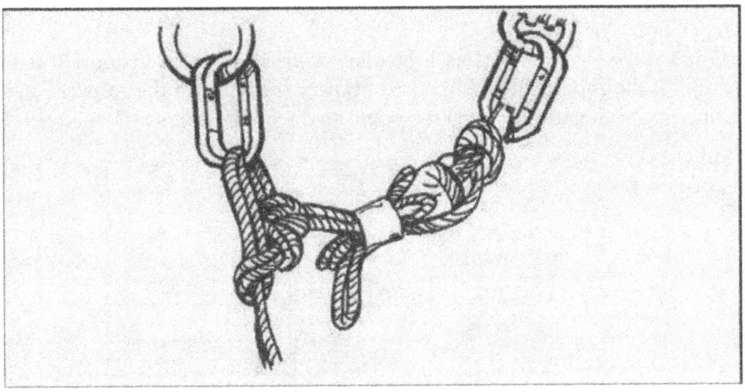

**Figure 3-16. Primary and secondary snaplink attaching points.**

h. Removes and secures the cargo hook access door and deploys the cargo hook in the DOWN position.

i. Installs the recovery rope for endangered rappellers.

j. Tapes the unused floor rings.

**Note:** There must be a minimum of 22 inches and a maximum of 30 inches between the anchor knots.

### 3-24. SEATING ARRANGEMENTS AND LOADING TECHNIQUES

Rappelling operations are executed quickly and safely to ensure the success of the operation in a combat or training environment. All rappellers must know the seating arrangements and the procedures for loading. They must be rehearsed under the supervision of the rappel master before conducting the operation.

a. A maximum of 10 rappellers, with and without combat equipment, and 1 static rappel master is seated and restrained aboard the UH-60 helicopter when the center row of troop seats are removed (Figure 3-17).

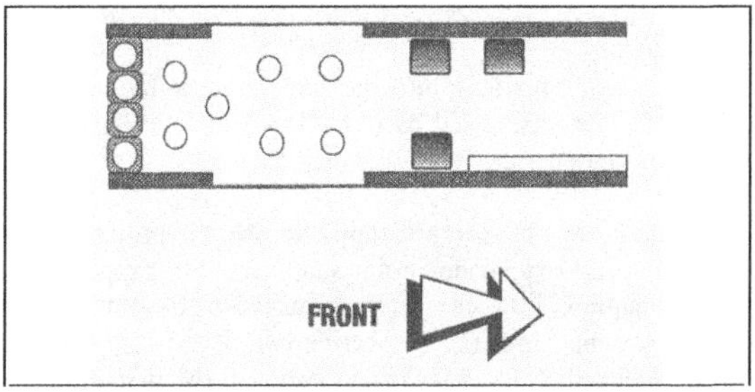


**Figure 3-17. Rappelling personnel seating arrangement.**

(1) Rappellers No. 1 through 6 and the rappel master are seated on the cargo floor. The rappellers are restrained using 3 1/2 feet of 1-inch tubular-nylon webbing (or equivalent) that is run through the cargo tie-down rings attached to the floor of the aircraft. The webbing is attached to the rappeller by means of a snaplink that is hooked to a Swiss seat. The rappel master is secured by his safety harness.

(2) Rappellers No. 7 through 10 are seated on the troop seats across the aft end of the cargo compartment. They are secured by the seat belts.

b. Rappellers are organized into 10-man sticks. They approach the aircraft from the left or right side in reverse order. They are divided into two groups and approach and enter the aircraft at the same time, depending on the situation. Rappellers No. 9, 7, 5, 3, and 1 enter the aircraft through the left door; they are seated and restrained. Rappellers No. 10, 8, 6, 4, and 2 enter the aircraft through the right door; they are seated and restrained. The static rappel master is restrained in the center of the cargo compartment using a C-3A troop-type safety belt or a safety harness during take-off and landing.

**DANGER**

**BECAUSE THE LOWEST ARC OF THE ROTOR BLADE OCCURS AT THE DIRECT FRONT OF THE AIRCRAFT, APPROACHING THE AIRCRAFT AT THIS POINT COULD RESULT IN PERSONNEL INJURY OR DEATH.**

### 3-25. RAPPELLING PROCEDURES

The rappel master ensures all personnel follow the rappelling procedures for the safe and efficient execution of the operation. These procedures are rehearsed under the supervision of the rappel master.

a. The rappel master is equipped with a headset and maintains communications with the pilot at all times. He wears a safety harness and stations himself in the center of the cargo floor to maintain control of the rappellers, their ropes, and all anchor attaching points.

**Note:** If all rappellers are rappelling into the same area, individual rappel ropes are not needed. The ropes used by rappellers No. 1 through 4 are used by the succeeding rappellers.

b. After all rappellers are in the aircraft, rappellers No. 1 through 4 hand the rappel master their primary and secondary anchor point snaplinks. The rappel master hooks the primary anchor point snaplink into the respective cabin tie-down fittings and the secondary anchor point snaplink into the respective cargo restraint net rings. After the primary and secondary anchor point snaplinks are secured, the first four rappellers hook onto their respective rappel rope. The rappel master checks the first four rappellers for the correct hookup.

c. After the first four rappellers exit and clear their ropes, the rappel master hands the next four rappellers their rappel ropes to hook up. Left-handed rappellers take extra precaution to ensure the correct hookup. Problems can be avoided by placing left-handed rappellers in the positions where they initially have the rope to their nonbrake side.

**WARNING**

**Extreme care must be taken in hooking up and positioning personnel that require repositioning of their rope to their proper brake-hand side. This could result in bodily injury or damage to Army property.**

d. After checking the rappellers for the correct hookup, the rappel master moves the next four rappellers into position on the floor of the aircraft. After the four rappellers exit and clear their ropes, the rappel master hands the last two rappellers their rappel ropes to hook up.

e. After checking the rappellers for correct hookup, the rappel master moves them into position on the floor of the aircraft. After the last rappeller is off the rope, the rappel master releases the ropes.

f. If the rappel master rappels, the crew chief is instructed to release or retrieve the ropes.

### 3-26. RAPPELLING COMMANDS

The UH-60 rappelling is conducted similar to the UH-1H rappelling with the exception that no skids are on the UH-60 on which to stand. Therefore, the edge of the floor, along the door of the helicopter, is used as a pivot point to assume the L-shaped rappelling position.

a. GET READY. This alerts the rappeller. The rappeller and rappel master perform final checks of the hookup, rappel seat, snaplink, and equipment.

b. THROW ROPES. The rappeller looks down below the aircraft to ensure that no one is under the aircraft. Keeping his brake hand in the small of his back, he tosses the deployment bag with the ropes *out and away* from the helicopter with his guide hand. The rappeller observes that the ropes are touching the ground and are not knotted or entangled.

c. POSITION. From a kneeling position with the brake hand in the small of the back, the rappeller rotates 90 degrees so that he is facing the inside of the aircraft and the rappel master. The rappeller then places his heels on the edge of the floor of the helicopter doorway and leans out into an L-shaped position. His feet are shoulder-width apart, the balls of his feet are on the edge of the helicopter doorway, his knees are locked, and his body is bent at the waist toward the helicopter.

d. GO. The rappel master initiates the rappel with this command. The rappeller flexes his knees and jumps backwards. At the same time, the rappeller throws his brake hand out at a 45-degree angle, letting the running ends of the ropes slide through both the brake hand and guide hand.

### 3-27. INSPECTION AND SAFETY CONSIDERATIONS

The rappel master and pilot (or his representative) conduct a joint inspection of the aircraft to ensure the safety of all personnel and serviceability of equipment.

a. Cargo doors are locked in the open position or cleared for closing, depending on the mission.

b. All loose objects in the cargo compartment are removed or secured forward.

c. Sharp edges or protrusions on the cargo floor and door ledges that may come in contact with the rappeller or his rappelling rope are taped.

**Note:** Do not tape the door latches or handles; this can interfere with door operations.

d. Primary and secondary rappelling anchor points are serviceable and securely attached to the aircraft structure.

e. A headset/helmet and intercom jack for the rappel master are available and operational, and the intercom extension cord is secured overhead.

f. Serviceable safety harnesses are available for the rappel master and crew chief.

g. The cabin ceiling tie-downs must have safety wire installed to ensure they do not come undone or unraveled, and the bolt head must be stamped H.

## Section VII. RAPPELLING OPERATIONS FOR MH-53 HELICOPTER

The MH-53 helicopter is a highly versatile aircraft. It is used for a variety of missions, usually in conjunction with special operations.

### 3-28. CHARACTERISTICS

The MH-53 is a two-engine, single-rotor, heavy-lift helicopter. It has a crew of six and can be refueled while in flight. It has a precision navigational and communications package, which makes it excellent for conducting deep infiltration and exfiltration missions. The MH-53 helicopter can fly at night in all weather and terrain conditions. It can follow down to 100 feet and operate from unprepared sites. It is fitted with three gun stations for a mix of 7.62-mm miniguns and caliber .50 machine guns.

### 3-29. RIGGING OF MH-53 HELICOPTER FOR RAPPELLING

When rappelling from the crew entrance door, rappellers connect the ropes with locking snaplinks to the two 10,000-pound tie-down rings under the left scanner's window. When rappelling from the ramp, rappellers connect the ropes with locking snaplinks to the two 10,000-pound tie-down fittings on the left and right sides of the aircraft at station 522. Any sharp edges that could damage the ropes should be padded or taped. A length of fire hose may also be used over the portion of the rope that comes in contacts with the door or ramp edge.

### 3-30. SEATING ARRANGEMENTS AND LOADING TECHNIQUES

The MH-53 is a large, versatile helicopter. The seating arrangements and loading techniques are numerous. The using unit develops a seating and loading SOP that ensures the safety of all personnel and permits smooth, efficient execution of the mission.

### 3-31. RAPPELLING PROCEDURES

The rappel master or his assistant are at the two exit points (ramp and door). The rappel master maintains communications with the helicopter commander and relays all commands and time warnings. The commander issues time warnings at 20-minute, 10-minute, 5-minute, and 1-minute intervals. During limited visibility, the rappel master may use NVDs to observe the safety/belayer. Chemical lights are attached to the top or bottom of the rappel rope. Arm-and-hand signals should be IAW The Air Assault School Handbook and FC 57-2.

**Note:** Night vision goggles cause limited depth perception and a tunnel vision effect.

a. The rappel master is secured by a harness and ensures the proper hook up of each rappeller. Once hooked up, the rappellers release their safety belts. At the command of the rappel master, the rappellers position themselves to ease immediate deployment.

**WARNING**

**Before conducting deployments off the ramp, the deploying team members must be briefed on the importance of maintaining separation between members. About 24 to 27 inches are desired. This separation helps maintain aircraft center of gravity limits.**

b. The rappel master deploys ropes only after the helicopter is in a stable hover over the target area and he has given the command ROPES. He ensures the ropes are on the ground before giving the command GO. After the last rappeller is off the rope, the rappel master retrieves or cuts the ropes away.

## 3-32. RAPPELLING COMMANDS

The helicopter rappelling commands for the MH-53 are as follows:

a. GET READY. This command is given as the helicopter approaches the rappel site. It alerts the rappellers of the approach to the site and signals them for a final equipment check.

b. POSITION. This command clears the first rappellers into position for deployment.

c. THROW ROPE. This command is given once the helicopter reaches a stable hover.

d. GO. The rappellers exit.

## 3-33. INSPECTION AND SAFETY CONSIDERATIONS

The rappel master briefs all personnel participating in the operation. He conducts hands-on inspection to ensure the safety of all personnel and serviceability of equipment.

a. Tie-down fittings are serviceable.

b. All sharp or protruding edges that may come in contact with the rappelling ropes are padded or taped.

c. All ropes are retrieved or cut away before forward movement of the helicopter.

d. Only three rappellers are deployed at a time—two from the ramp and one from the personnel door.

CHAPTER 4

# SPECIAL PATROL INFILTRATION/EXFILTRATION SYSTEM

*The special patrol infiltration/exfiltration system (SPIES) provides an excellent form of exfiltration over short distances. It is not recommended for infiltration because team members are exposed the entire time. Due to the nature of SPIES operations, a thorough briefing is required for all participants before the operation. Personnel being extracted must train extensively in the SPIES extraction before infiltration. For the other personnel involved, a complete preoperations briefing is held before the operation. This is crucial when additional assets are involved other than the extraction helicopter such as gunships, aerial observers, or artillery support.*

## Section I. SPIES MASTER DUTIES AND QUALIFICATIONS

The SPIES master holds the rank of sergeant or above (may be waived by the commanding officer). The commanding officer ensures that this qualification is entered on the soldier's training record.

### 4-1. QUALIFICATIONS

To qualify to be a SPIES master, the sergeant must participate in at least three SPIES operations (observe twice and execute SPIES master duties once under observation). For example, he hooks up the helicopter, assists in the preparation of an operation, and conducts a successful operation under the supervision of a qualified SPIES master. A SPIES master is conscientious and thorough, and he knows all aspects of a SPIES operation. He can give an effective pilot's brief, use the aircraft communications equipment, and understand aviation terminology.

### 4-2. DUTIES

The SPIES master is responsible for the safe conduct of the SPIES operation.

a. **Preflight Duties.**

(1) Inventories and inspects all SPIES equipment.

(2) Briefs the pilot and other concerned personnel about details of the operation, especially the extraction and dismounting procedures.

(3) Ensures that an ICS helmet and gunner's belt are available for the SPIES master's use. Uses a sling rope if no belt is available. Connects and checks the operation of the ICS to be used. (ICS communications is established between the SPIES master and pilots on all SPIES operations.)

(4) Attaches the SPIES rope to the helicopter IAW the guidance in this chapter.

(5) Ensures that nothing is adrift in the aircraft that may fall on a team member.

(6) Checks the location of the emergency axe. Places the axe where readily available, yet secure enough so as not to endanger the men on the SPIES rope. (Inspects the axe to ensure that it is sharp.)

b. **Extraction Duties.**

(1) On arrival at the team's estimated position, assists the pilot in determining the exact location of the team members.

(2) As the aircraft approaches the team's location, aids the pilot (using the clock system) in placing the aircraft directly above the team.

(3) Requests permission from the pilot to drop the SPIES rope when the aircraft is hovering above the team.

(4) Drops the rope, taking care to avoid striking team members on the ground.

(5) Notifies the pilot when the rope is down, and reports all altitude corrections to ensure that team members reach all SPIES attachment points.

(6) Watches for the "thumbs-up" signal from all team members.

(7) On receipt of the "thumbs-up" signal, advises the pilot that the team is ready for extraction and requests a vertical lift-off.

(8) Advises the pilot of the team's position, the location of any potential obstacles, and the avoidance of horizontal movement.

(9) If a team member becomes entangled with an obstacle during the extraction, notifies the pilot and requests that the vertical lift be stopped. If the situation is critical, prepares to cut the SPIES rope (the anchor point or cargo straps) after the team members are secured to the obstacle or on the ground.

(10) When positive that all obstructions are clear, advises the pilot to obtain a safe altitude (about 350 feet AGL for training purposes or as the situation dictates in combat) or to transition into forward flight.

(11) At frequent intervals during the flight, advises the pilot on the safety status of all team members. Maintains a constant visual watch on the team and checks security of the SPIES attachments often.

c. **Dismounting Duties.**

(1) On arrival at the dismounting area, informs the pilot as to the approximate distance of the lower rope end from the ground.

(2) Once the pilot starts the vertical descent, continually informs him as to the approximate distance of the lower rope end from the ground.

(3) Informs the pilot of any horizontal drift that occurs and any obstructions near the SPIES rope. Also informs the pilot of any oscillation that may occur.

(4) Informs the pilot when the rope is about 25 feet above the ground and again when it is 10 feet above the ground. Ensures that the rate of descent is slow enough to enable the team members to land and get out from under team members safely.

(5) Reports when the first man initially touches down, when the last team member starts to safely move away from under the helicopter, and when all team members are disconnected.

(6) On order of the pilot, either retrieves the SPIES rope into the helicopter or disconnects the SPIES rope and drops it to the ground. When using the UH-1H/UH-60 helicopter, the only way to retrieve the SPIES rope while in the air is by having an arranged recovery rope attached with a 16-foot sling rope. In some cases, the SPIES master joins two 12-foot-long sling ropes to haul the SPIES rope aboard and attaches the rope about 5 or 6 feet below the cargo hook or cargo strap hookup point. The type of knot used to connect the sling (or recovery) rope to the SPIES rope is self-tightening in nature (for example, the prusik knot). The SPIES master fastens the standing end of the sling rope to the deck tie-down or uses a snaplink. Although it is important to keep the line out of the way, the primary consideration is its length. The rope must be long enough for any oscillation in the SPIES during flight.

## Section II. PREOPERATIONS BRIEFINGS AND PROCEDURES

Before conducting a SPIES training mission, the participants must have a basic understanding of the requirements. The SPIES master conducts briefings to ensure the soldier extracted and the pilot know the procedures. As SOPs are developed and units train together, the SPIES master simply refers to the SOP. He always gives a safety briefing and conducts an equipment inspection.

**Note:** SPIES has not been approved for Army-wide use. Approval must be obtained from division level or higher commander before the conduct of any training.

### 4-3. SAFETY

As in all training, a safety briefing must precede operations using the SPIES. The briefing should consist of, but not be restricted to, a review of the following:

- Characteristics of equipment associated with the SPIES.
- Equipment inspection.
- Proper donning of the harness.
- Methods of extraction and insertion used.
- Emergency signals.

### 4-4. COMMUNICATIONS

Due to the noise level associated in all helicopter operations, radios are used for communication. They are also used before the arrival of the helicopter. Precise arm-and-hand signals are established in case of radio failure or poor communications due to static or noise overriding the audio output of the radio. During the first part of the operations, the SPIES master either observes (daytime) or knows that a definite procedure is taking place (night or jungle) while the teams are hooking up to the SPIES rope for extraction. Headsets and voice suppressors replace the handset for better radio procedures. This allows the RATELO on the ground to use both hands while the helicopter hovers for a faster and safer hookup. If radio communications are hampered, special procedures are used along with hand or light signals.

### 4-5. EXTRACTION PROCEDURES

After the team has been located, the SPIES master assists the pilot in directing the helicopter to the proper distance over the team.

a. The team leader positions himself so that he can move and approach the rope as it is dropped by the SPIES master. Once the rope is clear of any obstacles, the team leader signals the team to their assigned positions along the 10 hookup points. Using the primary or harness snaplink, each team member hooks to the D-ring on his side of the line. This is the primary hookup. Once the team is hooked up, the team leader hooks into the alternate or second hookup point using the safety line and snaplink. He then faces forward along the line so that he is heading in the direction he is traveling when the aircraft starts its ascent.

b. The SPIES rope is held up and routed over the shoulder closest to the rope. With the other hand, the team members give thumbs-up signals to allow both the team leader and the SPIES master to see that they are ready to go. The team leader physically inspects (if time and situation permits) or hooks himself in on the lowest point along with the RATELO to ensure the running end is clear of all obstacles. He gives the thumbs-up signal to the SPIES master. This thumbs-up signal (at night, an arranged light signal) continues until the team is lifted off the extraction surface. The team leader also signals (arms held straight to the sides) when the team is at safe altitude (approximate 10 feet above the tallest obstacle at the extraction site). The helicopter starts a transition in a horizontal direction on its return flight.

## 4-6. EMERGENCY PROCEDURES

During the flight, from extraction until the team is safely and quickly detached from the SPIES rope, each team member should be aware of any problems that arise from above or below. The man above checks the man below. At the first sign of danger or if an emergency arises, the team leader or a team member places his free hand on his head. Upon observing anyone on the SPIES rope with his hand on his head, the SPIES master tells the pilot to make an emergency landing in the nearest and safest area.

## 4-7. DISMOUNTING PROCEDURES

During the familiarization training phase, the SPIES master ensures all members know they should immediately head in the direction of the nose (12 o'clock) of the aircraft. This allows the pilot to see that the team is out from under the aircraft. When an emergency situation arises, the pilot can make a better appraisal of the situation if he can see all the members of the team. When the helicopter is making a scheduled landing, the team ensures that the SPIES rope does not interfere with the aircraft and that the aircraft does not land on the rope.

## 4-8. INSPECTION OF EQUIPMENT

The SPIES is inspected by a certified SPIES master at six-month intervals and whenever the SPIES master questions serviceability. Outdated, spliced, abraded, or cut rope is removed from service. However, loose or broken stitching in excess of three stitches is not repaired.

> **CAUTION**
> **Acid contamination, cuts, or fraying of harness or sling webbing constitute nonrepairable damage.**

a. The certified SPIES master inspects the harness and suspension sling webbing for signs of contamination from oil, grease, acid, or rust. He also checks for cuts, twists, fading, excessive wear, fusing (indicated by unusual hardening or softening of webbing fibers), fraying, burns, abrasions, and loose or broken stitching. The damaged harness or suspension sling is removed from service and returned to supply for appropriate disposition.

b. The SPIES master inspects all hardware for signs of corrosion, pitting, ease of operation, security of attachment, bends, dents, nicks, burrs, and sharp edges.

Replacement of hardware (except chest strap adapter) that requires unstitching of webbing renders the harness unserviceable.

c. The SPIES master replaces the V-ring by cutting the strap above the stitching. He folds and stitches a new end section of the leg strap. If the leg strap is damaged, the SPIES master returns the harness or suspension sling to supply for appropriate disposition.

d. The SPIES master checks ropes, harnesses, and suspension slings for expiration of service or total life, which is 7 years of service (opening manufacturer's package) or 15 years from date of manufacture, whichever occurs first. The SPIES master also ensures the rope is free of splices. He ensures the eye loop at the end of the SPIES rope is not broken, frayed, or loose.

e. The SPIES master inspects the rope surface for cuts, excessive abrasions, and snags. Cuts on the rope are excessive when there are four or more cut strands in any 5-inch length. The two-to-one braided rope has 12 pairs of strands (24 individual strands) around the circumference. Abrasion is extensive when torn yarns are equivalent to that of four strands of any 5-inch length. Rope that has been subjected to heavy loads may display glazed areas where it has worked against hard surfaces. This condition may be caused by paint or fused fibers. After long use, the rope may become fuzzy on the surface (although this should be minimized with the surface coating). However, the effect on the rope's strength is negligible.

f. The SPIES master inspects the rope for signs of contamination by acid, alkaline compounds, salt water, fire extinguishing solutions, and petroleum-based solvents. Although the ropes gradually change color, such changes do not indicate decrease in strength unless the change is due to contact with strong chemicals. Changes in color caused by chemicals are usually spotty; changes caused by use are uniform throughout the length of the rope.

## 4-9. RIGGING OF UH-1H HELICOPTER FOR SPIES OPERATION

Personnel who are not familiar with SPIES are encouraged to observe or take part in the rigging of the helicopter. This builds confidence in the equipment and assists in a more comprehensive training of new SPIES masters.

a. **Equipment.**

- One 120-foot SPIES rope with deployment bag.
- Two 11-foot, 3- or 4-loop cargo slings or two 9-foot, 3- or 4-loop cargo slings.
- Two Type IV connector links (four if aircraft does not have cargo hooks).
- Heavy duty tape (100-mph tape).
- One 12-foot sling rope.
- Five oval snaplinks (nine if the aircraft does not have cargo hooks).

**NOTE:** If the aircraft does not have cargo hooks, use four 11-foot or four 9-foot, 3- or 4-loop cargo slings.

b. **Installation.**

(1) The primary attachment point for the SPIES rope is the cargo hook. The end of the SPIES rope having a polyurethane encapsulated eye is attached to the cargo hook.

The two 9- or 11-foot long cargo suspension slings are joined to form one continuous sling, using a Type IV link. The sling is stretched out on the helicopter deck. One end is taken under the helicopter and through the eye of the SPIES rope. It is then connected on the other end of the sling using a Type IV link assembly. The sling must pass between the helicopter skids and the fuselage.

(2) Once the the SPIES rope and cargo straps are in place, the straps running across the deck of the helicopter are secured in place by at least four and as many as eight snaplinks. The snaplinks are evenly spaced across the deck and alternated from one side of the strap to the other and top and bottom, so that the first snaplink is to the rear of the strap and going around the bottom two straps. The next snaplink is in the front of the cargo strap and around the top two sections of the strap. This process continues until at least four points are established.

(3) If eight snaplinks are available, then each tie-down has two reversed. If no hook is available or is not working properly, the SPIES can be used safely by doubling the cargo slings and Type IV links. Thus, two cargo straps are side by side for a total of four slings and four Type IV links. Padding is used around the edge of the fuselage to protect the sling from damage.

(4) Care must be taken when using the UH-1H helicopter because of the ways in which it can be outfitted. Some helicopters have a step attached at the doorway. This is an added obstruction not only during installation but also during the operation. Others may have rocket pods or machine guns mounted. Not all of the UH-1 helicopters are hooked up the same way each time.

## 4-10. RIGGING OF UH-60 HELICOPTER FOR SPIES OPERATION

The SPIES master inspects all materials required to rig the helicopter.

a. **Equipment.**

- One 120-foot SPIES rope with deployment bag.
- Two 11-foot, 3- or 4-loop cargo slings or two 9-foot, 3- or 4-loop cargo slings.
- Two Type IV connector links (four if aircraft does not have cargo hooks).
- Heavy duty tape (100-mph tape).
- One 12-foot sling rope.
- Five oval snaplinks (nine if the aircraft does not have cargo hooks).

**Note:** If the aircraft does not have cargo hooks, use four 11-foot or four 9-foot, 3- or 4-loop cargo slings.

b. **Installation.**

(1) The primary attachment point for the SPIES rope is the cargo hook. The end of the SPIES rope has a polyurethane encapsulated eye that is attached to the cargo hook. The two 9- or 11-foot cargo suspension slings are joined to form one continuous sling using a Type IV link. The sling is stretched out on the helicopter deck. One end is taken under the helicopter and through the eye of the SPIES rope. It is then connected on the other end of the sling using a Type IV link assembly. The sling passes between the helicopter skids and the fuselage. Padding is used around the edge of the cargo hatch to protect the sling from damage.

(2) Once the SPIES rope and cargo straps are in place, the straps running across the deck of the helicopter are secured in place by at least four and as many as eight snaplinks. The snaplinks are spaced evenly across the deck and alternated from one side of the strap to the other and top and bottom. Thus, the first snaplink can be to the rear of the strap, wrapping around the bottom two straps. The next snaplink is in the front of the cargo strap and around the top two sections of the strap. This process continues until at least four points are established.

(3) If eight snaplinks are available, each tie-down has two snaplinks connecting the same spot, and the swing gates are reversed. If no hook is available or is not working properly, the SPIES can be used safely by doubling the cargo slings and Type IV links. Two cargo straps are side by side with a total of four slings and four Type IV links.

### 4-11. RIGGING OF CH-46/CH-47 HELICOPTER FOR SPIES OPERATION

The SPIES master inspects all the materials required to rig the helicopter.

a. **Equipment.**

- Two 11-foot, 3-loop slings.
- Two 9-foot, 3-loop cargo slings.
- Four Type IV connectors.
- One 13-foot sling rope.

b. **Installation.**

(1) The SPIES rope is attached using two 9- or 11-foot cargo suspension slings and four Type IV links. The cargo slings are passed through the encapsulated eye of the SPIES rope and attached to the outboard cargo tie-down rings on the aircraft floor. Two tie-down rings are used for each sling. Padding is used around the edge of the cargo hatch to protect slings from damage.

(2) Not all of the tie-down rings will be in the exact same position on the helicopters. This is one of the main considerations in deviating from the installation procedures. When possible, the cargo straps are placed to form two U shapes. One strap is placed forward of the cargo hole in the center of the aircraft floor and one aft or toward the rear of the helicopter. Installed properly, the cargo straps hold the SPIES rope comfortably in the center of and slightly below the opening of the cargo hatch.

(3) The use of snaplinks attached close to all four faulty tie-down points not only ensures a backup in case of a faulty tie-down ring but also reduces the amount of movement in the cargo suspension straps. A total of eight snaplinks are used for added security at each point with the swing gates reversed.

## Section III. LAND AND WATER EXTRACTION

If the situation, mission, or terrain suggests the possibility of a SPIES extraction, the leader includes the SPIES harness in each individual's equipment list. If the mission or insertion precludes the wearing of the harness, it will be carried on the top and just inside the flap of the pack being used. Once the extraction helicopter has been requested, the harness is retrieved and donned.

## 4-12. LAND EXTRACTION PROCEDURES

The SPIES is used only when the team requires immediate extraction or is unable to move to a clear (open) position suitable for helicopter landing.

a. The extraction helicopter proceeds to the area and establishes radio or visual contact with the team. The backup helicopter equipped with the SPIES remains aloft and away from the area. The helicopter maintains visual contact with the LZ and monitors radio communications.

b. After the team has indicated readiness for pickup and the tactical situation has stabilized, the extraction helicopter moves to the LZ by the safest route. When the helicopter is above the team's location, the SPIES master drops the SPIES rope on order of the pilot after the aircraft has obtained a stable hover at slightly above treetop height.

c. The team hooks up the same as in the familiarization procedures. Team members sling individual weapons over the shoulders, securing weapons and equipment to withstand the wind. A safety line is attached to prevent lost weapons during SPIES operations. When the team is ready, the team leader gives a thumbs-up signal.

d. During extraction, the team RATELO maintains communications with the extraction helicopter. He gives a verbal backup to the thumbs-up signal and relays all information during the flight. His location is near or at the bottom hookup point so he can assist in giving accurate information about the extraction, clearing of obstacles, and descent.

e. Lift-off of the extraction helicopter is vertical until the SPIES rope has cleared all obstacles. With the barrel directed downward at a 45-degree angle and outward, the team members fire their individual weapons using the hip position.

f. Once the SPIES rope is clear of all obstacles, the extraction helicopter changes to horizontal flight and departs the area by the safest route. Airspeed is limited to 70 knots in moderate climates and 50 knots in cold climates while team members are attached to the SPIES rope.

g. When the extraction helicopter has reached a safe dismount area, a transaction is made into hover flight at an altitude of 250 feet AGL. A vertical descent is started with the SPIES master continuously providing information to the pilot on the distance from the ground to the lower end of the SPIES rope. The vertical descent rate of the aircraft (at touchdown) is less than 5 feet per second.

h. When the team reaches the ground, they immediately move out toward the front of the helicopter. All team members rapidly unhook themselves and teammates that need assistance. Once this is completed, they either move away from the area and set up security or assist in clearing the rope if the helicopter is going to land.

## 4-13. WATER EXTRACTION PROCEDURES

The SPIES is suitable for extracting teams from the water. For this procedure, three inflatable life vests or any type of floatation device is tied to the SPIES rope to provide buoyancy for the rope while in the water. One floatation device is tied at each end of the attachment points, and one floatation device is tied in the middle of the attachment point area just above the middle two sets of D-rings. The team members to be extracted wear their SPIES harnesses under their individual life vest. They can also wear swimming fins, masks, and snorkels (amphibious operations) to ease hooking up to the SPIES rope within the spray area beneath the hovering helicopter.

a. After the extraction helicopter has attained a stable hover above the team member's location, the SPIES master drops the SPIES rope (with floatation attached) on order from the pilot. When the team members complete the hookup to the SPIES rope, the team leader signals the SPIES master to start lift-off. Lift-off of the extraction helicopter is vertical until all team members and the bottom end of the rope have cleared the water. During the initial lift-off, team members know that they are to be dragged through the water. The team members prepare to roll on their backs until clear of the water.

b. Flight speed and altitude are the same as over land. The dismounting procedures remain the same, except when landing on a ship. Once on board, the team members take their orders from personnel in charge of the deck.

## Section IV. AFTER-OPERATIONS PROCEDURES

After-operations procedures include repairing, cleaning, and proper storage of the rappelling equipment.

### 4-14. REPAIRING AND CLEANING OF EQUIPMENT

The contaminated ropes are washed with a mild detergent, such as liquid dish soap, and cold water followed by a rinse in clean, fresh water. Ropes are dried at a temperature not to exceed 140 degrees Fahrenheit. Stubborn oil, grease, hydraulic fluid, and other petroleum stains can be removed with the cleaning agent xylene (Grade A or B, TT-X 916).

### 4-15. STORAGE OF EQUIPMENT

To avoid ultraviolet deterioration, the nylon materials should be protected from direct sunlight. The SPIES rope is stowed in an aviator's kit bag for protection when not in use. Bins or similar facilities are used for storage of SPIES equipment. Shelves used for storage should be at least 4 inches from the walls and 12 inches from the floor. Areas used for storage should be well ventilated and free of oil, acid, cleaning compounds, and other contaminants. Equipment must *not* be stowed above or near hot water pipes or heating apparatus.

CHAPTER 5

# FAST-ROPE INSERTION AND EXTRACTION SYSTEM

*The fast-rope insertion and extraction system (FRIES) is employed by small units to make rapid infiltration and exfiltration using rotary-wing aircraft in confined areas. The use of FRIES is normally restricted to situations where aircraft cannot land, and when mission requirements preclude the need for large amounts of equipment, and ammunition or heavy crew-served weapons.*

**Note:** Headquarters, Department of the Army (HQDA) policy specifies that FRIES is not approved for Army-wide use and names the Commanding General, United States Army Special Operations Command (CG USASOC) as the executive agent for FRIES doctrine. The use of FRIES is restricted to special operations forces, pathfinders, long-range surveillance units, and HQDA-approved schools with a USASOC-approved FRIES program of instruction. Requests for other units to conduct FRIES operations should be submitted to Commander, USASOC, ATTN: AOOP-TRS, Fort Bragg, NC 28307-5000.

## Section I. GENERAL

FRIES information is subject to change. Units that recognize the need to update this chapter should submit recommendations to Commander, USASOC, ATTN: AOOP-TRS, Fort Bragg, NC 28307-5000.

### 5-1. OBJECTIVES

The objectives of this chapter are to—

- Prescribe the requirements for FRIES qualification and training, including proficiency sustainment, safety, and administrative procedures.
- Establish the scope of the duties and responsibilities of key personnel who conduct FRIES operations.

### 5-2. GUIDANCE FOR COMMANDERS

Units that have HQDA approval to perform FRIES operations are authorized to conduct initial FRIES qualification and FRIES master (FRM) qualification training. Training requirements are established in this publication, USASOC Regulation 350-6, and applicable SOAR policies. Aircraft material requirements are specified by the US Army Technology Applications Program Office (TAPO) .

a. FRIES training and operations are inherently risky; therefore, safety is paramount. All training and operations require in-depth attention to detail from planning through preparation and execution.

b. The following basic guidance applies to FRIES training and operations:

(1) Unit commanders must personally approve FRIES training sites.

(2) A risk analysis and or assessment must be completed before FRIES operations.

(3) Night FRIES operations are medium risk or higher.

(4) Units authorized to conduct FRIES operations may incorporate FRIES activities in field training exercises only after personnel have become FRIES qualified.

(5) Field exercises using FRIES activities are subject to the limitations imposed by applicable directives, including mandatory support requirements.

(6) Requests for waiver of procedures or restrictions prescribed herein must be submitted in writing with endorsement of the first general officer in the requesting unit=s chain of command to HQ USASOC.

## 5-3. TRAINING PREREQUISITES

> **CAUTION**
>
> **Trainers must ensure all personnel preparing to participate in FRIES training are tested *with equipment* to ensure they have sufficient upper body strength to safely perform the full scope of roping duties.**

Before allowing soldiers to participate in FRIES training and operations, the unit commander must make sure the soldiers are physically fit. Minimum screening standards that apply to each training participant include—

- Having passed the Army physical fitness test within the past six months.
- Having passed a medical examination within the past two years.
- Being free of any injury or physical condition that could cause a potential safety hazard during FRIES training.
- Demonstration of ability to perform a controlled descent from a height of 15 feet while carrying a 40-pound load (not including the basic duty uniform and combat boots).
- Demonstration of ability to hold a static position on a FRIES rope for 20 seconds, using hands and feet to lock in while carrying a 40-pound load.
- Being qualified and current to fly FRIES activities IAW HQ USASOC and the Special Operations Aviation Regiment (SOAR) Aircrew Training Program and applicable policies (for aircrews).

**Note:** The peacetime maximum soldier load will not exceed 50 pounds. This includes helmet, weapon, vest, web gear, and rucksack. Rucksack weight will not exceed 35 pounds.

## 5-4. PERSONAL EQUIPMENT REQUIRED

The minimum personal equipment required for FRIES training or operations includes—

- Heavy leather gloves.

- A helmet with a chin strap.
- Protective goggles.
- A long-sleeve shirt or jacket, long pants, and boots.
- Hearing protection and identification tags for helicopter operations.

## Section II. FRIES QUALIFICATION TRAINING

Each major unit (commanded by an O5 or higher) is responsible for conducting its own FRIES qualification and sustainment program. FRIES trainers will be FRIES master (FRM) qualified and current. Unit training should be tailored to unit and situation needs, however proponent requisites will not be reduced without written approval of the CG USASOC.

### 5-5. INITIAL FRIES QUALIFICATION TRAINING

Before participating in fast-rope operations, personnel are briefed on the FRIES and its purpose, capabilities, and limitations. The briefing also covers the duties and responsibilities of the pilot in command (PIC), safety, FRIES master (FRM), the assistant FRIES master (AFRM), and any ground assistants. Once the FRIES briefing is conducted, the remainder of the initial training is hands-on practice of the proper FRIES operational techniques.

a. Individuals are shown the proper techniques for boarding the aircraft, moving to the door, grasping and descending, locking-in, and clearing the rope. Individuals are also shown the proper wear, hook-up, and use of the STABO extraction harness.

b. After the demonstration, all ropers participate in a practice exercise in which they properly perform the following tasks using a tower and subsequently an aircraft. All ropers must demonstrate a minimum of six properly executed FRIES descents (three without equipment and three with equipment) and at least one successful lock-in.

(1) *Perform FRIES descent and lock-in from a tower at the 34-foot level.* Training should be progressive, starting from a tower at the 34-foot level without equipment and then with equipment. All ropers must complete a successful lock-in at this level. Ropers will not progress above the 34-foot level until after demonstrating the ability to stop descent, lock-in, and hold a stationary position for 20 seconds with equipment.

(2) *Perform FRIES descent and lock-in from a tower at the 60-foot level.* The recommended succession will be no less than three successful FRIES descents on the 60-foot tower, again starting without equipment, completing a lock-in, and finishing with equipment.

(3) *Perform FRIES descent from a helicopter at 60 feet or higher.* The recommended completion should be at least two helicopter descents from 60 feet or higher (one without equipment and one with equipment).

**WARNING**

**During initial qualification, students will not perform rapid exits or have more than three soldiers on a rope at any time. During advanced training, the limit is no more than five personnel with equipment on any rope at any time. In all cases (insertion or extraction), the maximum load allowed on any fast-rope system will not exceed 1,500 pounds.**

**CAUTION**

**All ropers should be limited to no more than a total of 10 roping events in a 24-hour period; no more than 6 events should be with equipment.**

### 5-6. FRIES PROFICIENCY SUSTAINMENT TRAINING

Commanders must make sure soldiers participating in FRIES operations receive sustainment training on equipment and procedures within 24 hours before the FRIES operation. Ropers who do not attend FRIES sustainment training will not be allowed to participate in FRIES operations. As a minimum, training will include a review of the following:

- Arm-and-hand signals.
- Individual equipment riggings.
- Aircraft familiarization.
- Safety procedures.
- Any rehearsals the FRM or commander deems necessary.

## Section III. FRIES MASTER SELECTION AND QUALIFICATION TRAINING

Selection and qualification of FRIES masters (FRMs) is a unit prerogative. Units will maintain records of qualification and proficiency. These records will accompany SM to new units when a permanent change of station occurs.

### 5-7. FRIES MASTER SELECTION

Units may train and certify soldiers as FRIES-qualified FRMs. Soldiers selected by units to qualify as FRMs should be the rank of E5 or higher and should have demonstrated leadership abilities with emphasis on maturity and decisiveness.

### 5-8. FRIES MASTER PREREQUISITES

Soldiers performing FRM, AFRM, and safety duties must have completed initial FRIES qualification training, be experienced in FRIES operations, have completed FRM training, and have executed at least 10 FRIES descents and 2 extractions. An FRM candidate's experience must include no less than five descents with combat load and at least one night descent with equipment from 60 feet or higher.

### 5-9. FRIES MASTER TRAINING AND CERTIFICATION

During FRM training and certification, FRM candidates must participate in three FRIES helicopter operations (twice observing another FRM and once executing FRM duties under observation of a current FRM) and be certified by either the parent unit or an authorized school. FRMs must be proficient in FRIES operations and must demonstrate proficiency at the following:

- Inspecting, preparing, and rigging all FRIES gear.
- Inspecting and preparing aircraft for FRIES operations.
- Controlling and coordinating actions of AFRMs, safeties, and ropers.
- Preparing and conducting both insertion and extraction operations.
- Conducting troop and pilot briefings.
- Giving arm-and-hand signals.
- Giving time warnings and associated commands.
- Deploying and retrieving FRIES ropes.
- Rigging and lowering equipment.

### 5-10. FRIES MASTER REFRESHER TRAINING

Units will conduct refresher training to maintain acquired skills. FRMs who have not participated in FRIES operations during the past six months will receive refresher training by a current FRM and serve as an AFRM before performing FRM duties. Refresher training for ropers consists of an FRM briefing and participation in FRIES training.

## Section IV. KEY PERSONNEL DUTIES AND RESPONSIBILITIES

The following personnel duties and responsibilities provide baseline requirements for the safe conduct of FRIES operations. Unit standing operating procedures (SOPs) may increase (but will not reduce) training safety requirements.

### 5-11. AIR MISSION COMMANDER

The air mission commander (AMC) provides overall command and control when conducting large, complex missions. The AMC is selected or appointed by the headquarters responsible for the overall mission. He ensures that all operators, aircrew members, and support elements synchronize their actions during the conduct of the mission and the FRIES operation. An AMC is not usually needed during small-unit qualification or proficiency training events when only one helicopter is used.

### 5-12. FRIES TRAINING OFFICER

Units that conduct FRIES training appoint a training officer in charge (OIC) or a noncommissioned officer in charge (NCOIC) to manage the training. The OIC is responsible for planning and coordinating various aspects of training to include—

- Coordinating all support activities, such as procuring FRIES equipment, aircraft,

training areas, medical support, and communication.
- Assigning qualified personnel to perform FRM, AFRM, ground safety, and medic duties.
- Adhering to procedures for planning, preparing, and executing the operation IAW this TC, USASOC Regulation 350-6, and the participating unit=s SOPs.
- Ensuring the FRMs, AFRMs, PICs, aircrews, safeties, and medics are briefed on the operation.
- Ensuring FRIES training events are not conducted on or over unsafe ground. (The ground must be free of loose material that could become airborne and cause an unsafe condition or obscure vision for aviators or ropers.)
- Ensuring a medic, an aid bag, a backboard, and a dedicated vehicle and driver are on site during all training.
- Ensuring positive communication with aircraft is maintained throughout training.

## 5-13. PILOT IN COMMAND

The aviation unit providing helicopter support for the FRIES training and operation appoints the PIC. The PIC is responsible for all aspects of the flight and ensures—

- Aircrew members are current and qualified to conduct FRIES operations
- Aircrew members are informed of and understand their responsibilities in fast-rope operations.
- Procedures for planning, preparation and execution are adhered to IAW this TC, USASOC Regulation 350-6, and applicable SOPs and policies.
- Aircraft is properly configured to perform the mission.
- Aircraft is at the proper altitude, airspeed, and location as briefed.
- The command to deploy ropes is not given until the aircraft is at a stabilized hover.
- The aircraft position is maintained to keep ropes in contact with the surface until all descending ropers are on the ground (or extracting ropers are securely attached for extraction).
- Ropes are fully recovered inside the aircraft (or jettisoned) before the aircraft departs the stabilized hover position at the infiltration site.
- All personnel are briefed on in-flight emergency and safety procedures.
- Ropes are never deployed with anything other than night illumination attached to the free end.
- All personnel are briefed on fire support to be provided by the aircraft including:
  — The nature of fire support the aircraft can provide.
  — The time the fire support starts, shifts, and stops.
  — The primary and alternate commands and signals that start, stop, lift, or shift fires.

## 5-14. FRIES MASTER

Units conducting FRIES operations will designate one overall FRM to organize, coordinate, and supervise the activities of the day, and AFRMs as needed. AFRMs are additional FRM-qualified soldiers who assist the primary FRM. An AFRM will be designated for each roping point. AFRMs are responsible for all preparation, inspection, and command and control of all roping activities on their points. FRMs also must—

- Ensure all FRIES roping personnel, including FRMs, AFRMs, and safeties, are qualified and current and understand the proper roping procedures for FRIES operations.
- Perform safety and serviceability checks on all FRIES equipment and rigging.
- Ensure the attachment bar or points are serviceable and free of any defects or contamination and that the quick-release mechanisms and safety pins are present and serviceable and operate correctly.
- Inspect the rope to make sure it has no contamination, damage, or defects that could make it unsafe.
- Check the rope to make sure it is the correct type and length for the operation (smooth rope for infiltration or looped for exfiltration).
- Ensure the rope is properly attached with safety pin in place and back coiled.
- Ensure the extraction harnesses are serviceable and properly worn.
- Ensure the rope chemlights are correctly rigged and illuminated when needed (two at the mount, two at the end, and two 15 feet from the end).
- Confirm the proper seating arrangement, including seat belts or a safety strap for ropers.
- Coordinate with the PIC and brief the aircrew, safeties, and ropers on the following:
  - The aircraft flight, approach, flare, and position over the target site.
  - The height of the aircraft above the target.
  - A description of the FRIES site (rooftop, small clearing) and size, as well as the actions of the ropers upon arrival at the site.
  - The signal to deploy the rope, who will deploy it, and how it will be done.
  - Time warnings and actions at each time warning.
  - Arm-and-hand signals.
  - The use of radios and intercoms.
- Brief the PIC and crew chief on verbal commands and final adjustments over the target.
- Confirm the correct target and inform the pilot before the deployment of ropes or ropers.

**WARNING**

**The FRM must ensure that nothing, other than rope illumination, is ever attached to the free end of the FRIES rope during deployment and that equipment is never attached and dropped.**

- Deploy the ropes when the PIC gives the command DEPLOY ROPES. (This action may be delegated to the aircrew or safety.)
- Ensure safeties and AFRMS understand the time and activity sequence and all arm-and-hand signals.
- Ensure safeties know the proper procedures to—
- Keep the aircraft on target.
  - — Clear the ropes.
  - — Recover the ropes.
  - — Jettison the ropes.
- Observe the target and pass to the PIC, through the crew chief, final aircraft position adjustments over the target to make sure conditions are safe.
- Deploy the ropes when the PIC gives the command DEPLOY ROPES. (This action may be delegated to the safety.)
- Ensure all ropers know how to hook up for extraction and know the correct arm-and-hand signals.
- Direct the aircrew or safety to inform the pilot when the aircraft is on target (the PIC replies with the command DEPLOY ROPES).

**CAUTION**

**When equipment is to be lowered, it should be lowered before the FRIES is deployed or personnel begin roping. When simultaneous, equipment and ropers will exit from opposite doors of the aircraft.**

- — The safety repeats, " DEPLOY ROPES." The FRM signals the AFRMs to deploy ropes and makes a final check, ensuring all ropes from his aircraft reach the surface and are safe for descent (at least 15 feet of rope must be on the surface).
- — When conditions are safe, the FRM commands "GO" to the ropers and points to the rope or leads the stick out.

**Note:** If the FRM leads the stick out, the AFRM automatically assumes control. He completes marshaling the remaining ropers and is the last man to exit.

- Coordinate actions with the crew chief for the positioning or storage of seat belts, safety straps, rucksacks, door bundles, and communication cords.

**Note:** During night operations, upon completing descent, the AFRM observes the the ropes of his aircraft. When he sees that all ropes are clear, he issues the prearranged ALL CLEAR hand signal to the aircrew safety. This hand signal tells the aircrew that they are free to jettison or recover the ropes, and fly away.

## 5-15. FRIES AIRCREW MEMBER AND SAFETY

FRIES safeties are aircrew members who are FRM-qualified and perform the additional duties of the safety during helicopter roping operations. Safeties assist FRMs and pilots as needed. Aircrewmen performing safety duties differ from FRMs and AFRMs in that safeties are members of the aircrew (crew chief or gunner) whose orientation is the flying portion of the operation. Conversely, AFRMs are FRM-qualified members of the ground unit who are performing FRM duties as assistants to the primary FRM and who exit with the ground unit. Safeties remain in the aircraft and operate the doors, extend and retract the rope mount bars, and wear night vision goggles (NVGs) when appropriate. Safeties must have completed FRM training IAW this TC and the aircrew training program (ATP).

a. During roping operations, the safety's four major areas of responsibility are:

(1) Observe all activities in the aircraft and assist the pilot and FRM or AFRM as needed.

(2) Serve as the communication link between the PIC and the FRM. The safety monitors aircraft communication and relays information between the PIC and the FRM, constantly keeping both informed.

(3) Assist the FRMs and AFRMs in lowering cargo and clearing the aircraft of ropers.

(4) Clear and jettison the ropes, when appropriate.

b. The safety performs the following duties:

(1) Relays time warnings, aircraft positioning commands, and other information to assist the FRM.

(2) Conducts aircraft equipment rigging and serviceability inspections, making sure mounts, doors, and ropes are ready and safe for operations both pre-flight and en route.

(3) Verifies that the aircraft is at a stabilized hover with ropes in contact with the surface and remains so (or stops stick until corrections are made).

(4) Checks the rope to make sure 15 feet of rope are on the target surface.

(5) Tells the PIC, "ROPES DEPLOYED", when appropriate.

(6) Tells the PIC, "ROPERS OUT", when the first roper has exited.

(7) Helps the AFRM in controlling and spacing each roper's exit.

(8) Monitors the rope and ropers to make sure they do not get hung up, dragged, or dropped off site.

(9) Observes the exit of the last roper until the roper reaches the surface, moves clear of the rope, and gives the ALL CLEAR signal.

(10) Tells the PIC, "ROPERS DOWN", (station name: aft, forward, left, right) when all ropers are on the surface and conditions are safe to jettison the ropes.

(11) Passes the signals as required to position the aircraft, including HOLD, MOVE LEFT, RIGHT, FORWARD, BACK, UP, or DOWN.

(12) Acts as belay man during cargo lowering when needed.

(13) Jettisons ropes upon receiving the command JETTISON ROPES from the PIC. Tells the PIC, "Ropes are jettisoned," when the ropes have been recovered or jettisoned.

(14) Tells the PIC when ALL CLEAR.

**CAUTION**

**This command tells the pilot the aircraft is cleared for flight.**

(15) Deploys the rope for extraction upon receiving the PIC command DEPLOY ROPES.

(16) Observes the extracting rope and personnel and informs or guides the PIC through aircraft positioning and throughout the recovery flight.

**5-16. FRIES ROPER**

Ropers are responsible for notifying the FRM, AFRM, safety, or pilot if they observe any unsafe act or condition. Ropers may halt or call for a halt of roping for safety at any time. During FRIES insertion training, the number of ropers on the fast rope at one time is limited to three. During extraction training, the total weight per extraction bar or rope will not exceed a total of 1,500 pounds. When rucksacks are worn by ropers, the rucksacks must not exceed a weight of 35 pounds (total roper equipment weight will not exceed 50 pounds). Ropers will—

- Keep hands at head level.
- Maintain visual contact with lower ropers during their descent.
- Maintain a minimum of a 1-second interval on exit to avoid collisions.
- Keep at least two points of contact on the rope (both hands) at all times.
- Use their feet for additional breaking any time needed.
- Execute descents at a safe speed.
- Slow the rate of descent halfway down the rope to avoid landing on each other.
- Move quickly away from the rope upon arrival on the surface.
- Know correct wear of extraction (STABO) harness and extraction procedures.

**WARNING**

**The FRM, AFRM, and safety must make sure all personnel are on the surface and have moved clear of the ropes before aircraft fly away or ropes are jettisoned.**

## Section V. FRIES RIGGING OF AIRCRAFT

The aviation unit is responsible for preattachment inspection and installation of FRIES anchor point mount bars and associated equipment. All aircraft used to conduct FRIES operations with US Army personnel must be rigged and equipped IAW current parent service directives--for Army aircraft, USASOC, SOAR, and TAPO directives apply. Rigging of the aircraft includes installation and removal of the seats and seat belts (or personnel restraints), and attaching and detaching the ropes.

**Figure 5-1. UH-60 helicopter**

> **CAUTION**
> **Army aircraft that have the FRIES mounts installed must have an airworthiness release for FRIES operations from the Aviation Troop Support Command (ATCOM) TAPO.**

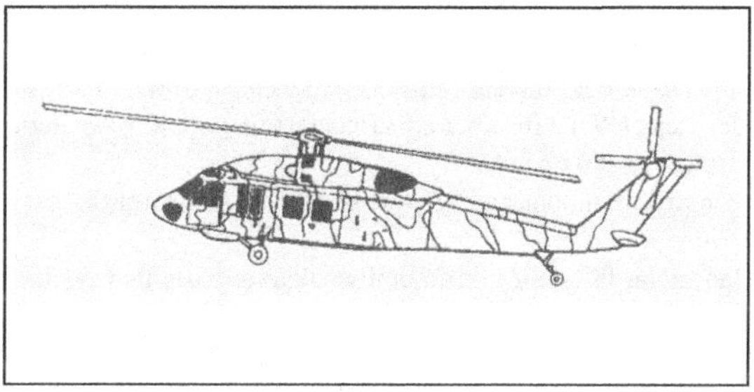

## 5-17. RIGGING OF FRIES IN UH/MH-60

The UH-60 is a twin-turbine, medium-speed, single-main-rotor helicopter (Figure 5-1).

The aviation aircrew or safety is responsible for—

- Inspecting the FRIES mount bar assembly, attachment hardware, and associated equipment to make sure they are serviceable and ready to use.
- Ensuring the bar assembly is correctly mounted to the aircraft (Figure 5-2).

**Figure 5-2. Bar assembly correctly mounted to aircraft.**

- Ensuring the mount is complete, functional, and free of any defect.
- Ensuring the mount and the aircraft floor are clean and free of any petroleum, oils, or lubricants (POL) or other contaminants that could weaken the rope or cause personnel to slip or fall.
- Ensuring the mount and aircraft are free of sharp edges or burrs that could cut the rope.
- Ensuring no metal fatigue or structural weakness exists that could cause the system to fail.
- Extending locking and retracting mount bars as needed.
- Lowering and belaying equipment as needed.
- Opening, locking, and closing cargo doors as needed.

**Note:** For cold-weather operations or during flights of long duration, the cargo doors are closed and latched until the time specified for opening. The aircrew will operate doors or directly supervises those who do. In good weather, doors should remain locked open.

- Removing or reconfiguring troop seats.
- Providing seat belts or floor-mounted personnel restraints for ropers' use during flight.
- Attaching and detaching ropes as needed.

**Note:** Clusters of rucksacks or small door bundles (weight limited to 250 pounds) may be lowered (belayed) on rappelling ropes. Cargo lowered this way must never be free dropped (the belay must start at the time the bundle is pushed [tipped] out the door).

## 5-18. RIGGING OF FRIES IN CH/MH-47

The CH-47 is a tandem-rotor, medium-transport helicopter (Figure 5-3).

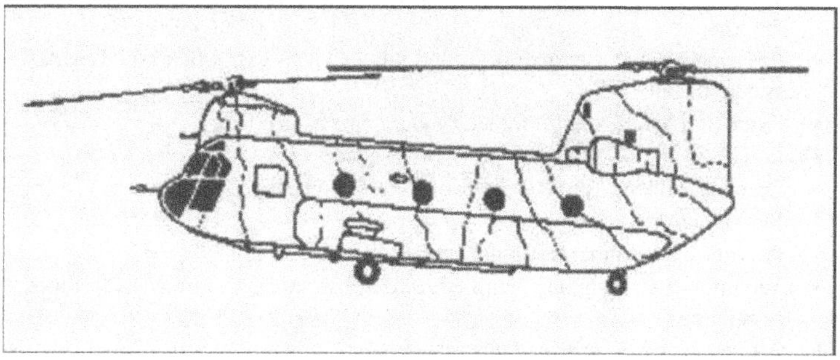

Figure 5-3. CH-47 helicopter.

The aviation aircrew or safety is responsible for:

- Inspecting the FRIES mount bar assembly, attachment hardware, and associated equipment to make sure they are serviceable and ready to use.
- Ensuring the bar mounts are correctly mounted to the aircraft (Figure 5-4, rear; Figure 5-5, winch).

Figure 5-4. Bar mounts correctly mounted, rear.

Figure 5-5. Bar mounts correctly mounted, winch.

- Ensuring the mount is complete, functional, and free of any defect.

- Ensuring the mount and the aircraft floor are clean and free of any POL or other contaminants that could weaken the rope or cause personnel to slip and fall.
- Ensuring the mount and aircraft are free of sharp edges or burrs that could cut the rope.
- Ensuring no metal fatigue or structural weakness exists that could cause the system to fail.
- Extending locking and retracting mount bars as needed.
- Lowering and belaying equipment as needed.
- Opening, locking, and closing cabin doors as needed.
- Removing or reconfiguring troop seats.
- Providing personnel restraints for ropers' use during flight.
- Taping and padding any sharp edges that can come in contact with ropes or personnel
- Attaching and detaching ropes as needed.

## 5-19. RIGGING OF FRIES IN MH-53

The MH-53 is a USASOC twin-engine, single-rotor, medium-transport helicopter used by the Army (Figure 5-6).

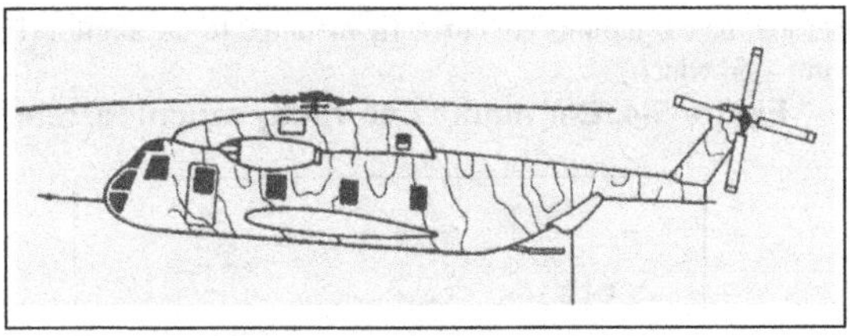

**Figure 5-6. MH-53 helicopter.**

The MH-53 primary FRIES mount bar assembly is located overhead, just aft of the rear ramp. The ramp mount bar assembly spans the cabin width and has a two-pin release mechanism on each end. The ramp position mount assembly provides two points; each will accommodate one FRIES rope or one cargo lowering rope. The ropes are carried back-coiled on the ramp.

a. The alternate position for fast-rope descent from the MH-53 is the crew entrance in the right side of the fuselage immediately behind the pilot's position. In this position, the fast-rope mount must be connected to the external hoist point. The FRIES master is responsible for inspecting the MH-53 mount system for its presence, security, and operability.

b. The aviation unit is responsible for—

- Installation and maintenance of the fast-rope mount assemblies at each position to be used.

- Configuring the aircraft to accommodate the number of soldiers and type of mission planned.
- Ensuring seat belts or restraint straps are present to secure ropers during flight.

**Note:** The MH-53 aircrew can improvise handrails to assist roper exits. When used, improvised handrails will be rigged IAW USAF special operations SOPs.

- Configuring the MH-53 for fast-roping when equipped with external 650-gallon fuel tanks.
  - Remove the ramp door to install the cross bar.
  - Place the bar in the brackets with the two quick-release paddles facing forward.
- Configuring aircraft armament (50-caliber machine gun and the GAU-2B ramp minigun) for FRIES operations.
  - When using the aft right roping position, point the 50-caliber machine gun to the 6 o'clock position and secure the gun with bungee cords.
  - When using the left aft rope, disconnect the minigun ammunition chute and stow it in the ammunition can, or disconnect it at the ammunition can and securely wrap it around the minigun. Use bungee cords to secure the ramp minigun in the firing position facing the 6 o'clock position.

**Note:** Ideally, only the right aft rope is used when the GAU-2B ramp minigun is installed. Although not recommended, the left aft rope may be used, if required.

## Section VI. EQUIPMENT MAINTENANCE AND INSPECTION

Before conducting a FRIES operation, the FRM must inspect all FRIES equipment (ropes, STABO harnesses, and mount bars) for serviceability and readiness for use.

**WARNING**

**FRIES ropes and harnesses are critical life support equipment and must be stored, cared for, and maintained as life support items.**

### 5-20. ROPES AND HARNESSES

FRIES equipment should be maintained in the same manner as a parachute. The unit rigger section can provide detailed guidance on appropriate inspection, care, and maintenance of FRIES equipment.

a. FRIES ropes and harnesses must always be stored in a clean, cool, dry space out of direct sunlight and free of chemicals or chemical vapors. Equipment that becomes wet with fresh water should be hung up to dry (indoors) on hardwood pegs. Equipment that is exposed to salt water or becomes imbedded with dirt or mud should be washed and rinsed in fresh water (within 72 hours) and hung up to dry (indoors) on hardwood pegs (out of direct sunlight).

b. Before conducting a fast-rope operation, the FRM—

- Inspects the fast rope thoroughly and carefully. Checks the rope length to ensure it is the correct rope for the operation planned.
- Checks the woven loop on the mount end for excessive wear or chemical contamination. Checks the rope along its entire length for fraying, cuts, and chemical contamination.
  - — Do not use a rope that is severely frayed. (Light fraying on the rope from normal use does not weaken the rope.)
  - — Do not use a rope when any single strand is cut halfway through or has two or more cuts that penetrate one-third or more through any strand's thickness within 1 foot of the running length of the FRIES.
- Inspects the rope for contamination of acid, alkaline compounds, saltwater, fire extinguisher solutions, or petroleum-based solvents. Changes in color caused by chemicals are usually blotchy and have an unusual odor. Although used ropes gradually change color, such changes do not indicate a decrease in strength unless the change is due to contact with strong chemicals. Changes occurring because of use are usually uniform throughout the length of the rope.

**WARNING**

**Any rope known or suspected to be chemically contaminated, cut beyond allowable limits, excessively frayed, or sun damaged must be taken out of service, marked for destruction, and turned in to the rigger section.**

- Inspects the extraction loops to the same standard as the main rope. Ensures the woven attachment loops are secure.
- Inspects the harness to ensure—
  - — Ropers are wearing the harness under all load-carrying equipment.
  - — Ropers have properly fastened all connectors.
  - — Harness material and stitching are not cut, torn, or contaminated, and all hardware is free of corrosion and is in operable condition.

**Note:** Serviceability inspection and maintenance of the extraction harness are the responsibility of the rigger-qualified personnel.

### 5-21. FRIES MOUNT BARS

The aviation unit is responsible for the installation, removal, storage, and maintenance of FRIES mount bars. The US Army Special Operations Aviation Regiment, Ft Campbell, KY, prepares and provides checklists for these procedures. Aviation units participating in FRIES operations must have and comply with current SOAR guidance.

Aviation units:

- Install and maintain the FRIES mount bar; check mount bolt torque and witness marks.
- Check the FRIES mount bar for security and operability.
- Check the condition of the mount bar; inspect for bends, dents, distortion, cracks, corrosion, and contaminants.
- Ensure the fast-rope quick-release pins are installed and operable (two each side).
- Check the security and condition of the rope-release system, as follows:
  - — The mounting bolts are installed and properly torqued.
  - — The safety pins are installed, and lanyards are present.
  - — The system locks and unlocks as designed.
- Ensure the release cables are attached at both ends and are not broken or kinked.
- Check the release handles to ensure they are securely mounted and can move freely.
- Check the latch as follows:
  - — Latch or close the release mechanism and apply a downward pressure on the release arm. Attempt to pull the release handles the release should not open.
  - — Relax pressure on the release mechanism and pull the handles the release should open.
  - — With no pressure on the release mechanism and with the safety pin installed, attempt to pull the release handles--the release should not open.

## Section VII. OPERATIONAL REQUIREMENTS AND LIMITATIONS

This section establishes medical coverage and communication support requirements for FRIES operations, including during adverse weather conditions and night operations.

### 5-22. MEDICAL SUPPORT

Dedicated medical support is required for FRIES training and demonstrations. A qualified medic with assistant, aid bags, two backboards, two stretchers, and a dedicated vehicle with a driver must be present to respond to any mishaps. The nearest medical evacuation (MEDEVAC) unit will be notified of where and when FRIES training is scheduled. MEDEVAC response will be coordinated as on call. Training or demonstrations will not be conducted without medical support present. The ground force commander may direct the FRIES aircraft to transport life or death (urgent) patients when no other MEDEVAC unit is available, providing the medical team is trained and properly equipped with the appropriate equipment for aerial transport of patients. This must be included in the air mission briefing prior to training and coordinated with the medical team providing coverage. FRIES operations that are evaluated as high risk or that occur at night must have a certified combat lifesaver on location and medics (equipped as stated above) in the immediate vicinity.

### 5-23. COMMUNICATIONS REQUIREMENTS

During peacetime training and operations, FRIES lift aircraft and ground elements will maintain continuous radio communications. Aircrew personnel clearing ropes at each FRIES station must be on the intercom system, monitoring aircraft internal communications. They must also be proficient in the use of arm-and-hand signals as precoordinated between FRIES users and aircraft crews. The designated aircrew member will relay time warnings between the PIC and FRM. Routine time warnings are 10-, 6- and 1-minute warnings to the FRM. Ground-to-air communications must be continuously maintained throughout FRIES training.

## 5-24. ADVERSE WEATHER OR TERRAIN CONDITIONS

FRIES training and operations will ***not*** be conducted under the following conditions:

- Anytime temperatures are below 32 degrees Fahrenheit or wind-chill factors that could cause cold-weather injuries exist including those caused by helicopter rotor wash or extraction cruise speed.
- Anytime water or ice on the rope prevents personnel from controlling their descent.
- Anytime the rope is exposed to freezing conditions for a sufficient length of time to freeze, thereby reducing its tensile strength.
- Anytime dust out or white out conditions restrict the aviators or the ropers vision.
- Anytime trees or vegetation on the LZ is too dense for safe training.
- Anytime the LZ is too steep or too slippery for safe training.
- Anytime winds exceed 30 knots or are too gusty for safe operations.

## 5-25. NIGHT OPERATION REQUIREMENTS

Night operations require the following procedures.

a. Attach two chemlights at the free end of the rope and two above the extraction attachment loops (15 feet from the end of the rope) to aid in determining when adequate rope is on the surface.

b. Attach two chemlights (one primary and one backup) to the loop of the rope, or to the rope mount, to guide personnel exiting the aircraft to prevent glare from obscuring vision.

c. Attach one chemlight or luminous tape to the top of each roper's helmet. If a chemlight is used, select a color different from the colors used on the ropes.

**CAUTION**

**Ropers will not wear NVGs during descent. NVGs limit depth**

**perception and cause a tunnel-vision effect. Static FRM and Safeties may wear NVGs to help confirm that fast ropes are fully deployed. The aircrew will wear NVGs as required by flight mission needs.**

## Section VIII. FRIES PROCEDURES

The risks of personal injury and equipment damage during FRIES operations can be reduced by detailed planning, thorough training, and strict adherence to safe procedures. Conducting FRIES operations with heavy loads requires personnel to be proficient in these operations. FRIES users must limit ropers to no more than 6 descents with combat load and no more than 10 total descents in any 24-hour period. The total weight of equipment soldiers may carry during training is limited to never more than 50 pounds. This section provides standardized procedures for FRMs, PICs, and ropers. These procedures are common to all units and aircraft, except as noted.

### 5-26. FRIES MASTER

The FRM performs the following series of actions.

- Briefs members of his team and aircrew.
- Inspects team members and equipment.
- Installs the FRIES rope in the aircraft and conducts safety checks.
- Relays time warnings to team members. (Time warnings are a tool to help keep aircrew and ropers' actions synchronized and can be modified according to user needs; however the 1-minute warning should always be used.)
- Breaks chemlights, if required. Two chemlights or luminous tape will be attached at the anchor point to identify the handhold point.

**WARNING**

**Equipment must never be attached to the FRIES and dropped. As little as 10 pounds of weight can exceed the load limit of the FRIES mounts when the load stops at the end of a 60-foot drop. The only items to be attached to the fast rope during deployment are the position markers.**

- Ensures the rope is properly configured for deployment (back coiled to prevent tangles).
- Ensures the team members are in order of exit no later than the 1-minute warning.
- Confirms target on final approach.
- Deploys the rope and ensures it reaches the ground. During night

operations, the rope is marked with six chemlights: two at the mount, two on the end, and two 15 feet from the end.

- Deploys personnel.
- Accounts for personnel, and signals aircrew.

> **WARNING**
> **Because the FRIES mount in the CH-47 helicopter is narrow, special care must be taken by ropers to avoid fouling weapons or equipment on the frame during exit.**

## 5-27. PILOT IN COMMAND

The PIC briefs the FRM, safety, and roping personnel before FRIES operations.

a. The PIC briefing includes:

- Approaching, loading, unloading, and departing the aircraft.
- Actions in the aircraft.
- Flight route and checkpoints en route.
- Altitude or height for roping.
- Time warnings to FRM as needed (recommend 10 minutes, 6 minutes, and 1 minute).

b. When the PIC is over the intended objective he calls, "DEPLOY ROPES" Fast-rope warnings and expected aircrew coordination calls are as follows:

(1) The aircrew or FRM provides any final position adjustments of the aircraft with the commands of MOVE (left, right, forward, back, up, down, or hold) as needed to place the ropes ON TARGET. (When the pilot is to hold the position, an explanation will be provided as soon as possible.)

(2) When the FRM deploys the ropes, the aircrew member calls, "ROPES DEPLOYED."

(3) When the first roper exits the aircraft, the aircrew member calls, "ROPERS OUT."

(4) The safety watches the last roper until he is on the ground and gives the prearranged signal. The safety then tells the PIC, "ALL ROPERS AWAY," and identifies his station. This alerts the PIC that the safety is going to either jettison or recover the ropes.

(5) The safety will then either recover or jettison the rope as planned, clear the aircraft all around for obstructions, and call, "ROPES CLEAR (station)" This informs the PIC that the aircraft is clear for flight.

c. The PIC also briefs on emergencies that could occur in the following situations:

- En route (when ropers are inside the aircraft).
- Before deployment of ropes.

- When fast-rope personnel are on the ropes.
- After the ropers are on the surface.

## 5-28. ROPERS

Ropers perform the following actions:

a. At the command STAND BY (given at 1-minute warning), ropers make final check of themselves and take the prepared-to-exit position.

b. At the command GO, the first roper exits the aircraft, rotates his body 90 to 180 degrees to make sure his equipment clears the aircraft, places the fast rope between the arches of his feet, and begins descent. The roper uses his hands and feet to slow his descent approximately two-thirds of the way down.

c. Subsequent ropers exit at 1-second intervals using the same procedure, but begin slow descent approximately half-way down to avoid landing on other ropers.

d. During descent, ropers keep a lookout and break as necessary to avoid landing on obstructions or on fellow ropers.

e. Ropers prepare to land just before reaching the ground by spreading their legs about shoulder-width apart with their knees slightly bent.

f. At landing, ropers quickly move clear of the ropes to avoid collisions with descending ropers.

**CAUTION**

**According to AR 95-3, anytime operations occur over water, all personnel will wear an approved personal flotation device as identified in FM 57-220 or USASOC Regulation 350-2.**

## 5-29. EQUIPMENT-LOWERING PROCEDURES

Anytime a roper's load exceeds a total of 50 pounds or his rucksack weight exceeds 35 pounds, equipment will be lowered (as clusters or bundles) on separate ropes. Lowering equipment from heights great enough to cause equipment damage necessitates belaying the equipment during descent to avert damage. Belaying devices used in mountaineering allow ropers or safeties to control equipment descent.

**Figure 5-7. Rucksacks upright on cargo deck.**

**CAUTION**

**When equipment is to be lowered, it should be lowered before the FRIES is deployed or personnel begin roping. When done**

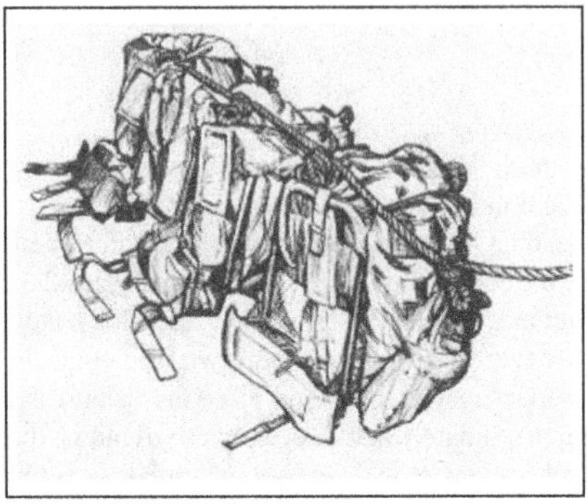

**simultaneously, equipment and ropers will exit from opposite doors of the aircraft.**

a. Rigging equipment to be lowered requires rigging rucksacks in clusters. Clusters of up to five rucksacks may be lowered (smaller clusters work better). FRMs will ensure the total weight of any cluster does not exceed 250 pounds.

(1) Place the rucksacks upright on the cargo deck of the helicopter (Figure 5-7) .

(2) Lay a sling rope across the frames between the shoulder strap attachment points.

(3) Tie a bowline knot at each end of the sling and a figure eight knot at each rucksack.

(4) Attach a snaplink to the frame of each rucksack, to the loops in the sling rope, and to the end of the lowering rope.

(5) For Lowe rucks, install the issue H-harness on the rucks and join them into clusters with a sling rope tied to the quick-release attachment buckles.

b. Rigging of the lowering rope to belay the rucksacks requires the following actions.

(1) Backcoil a rappel rope on the helicopter floor at the gunner=s position.

(2) Attach a snaplink and belay device (either a figure eight [Figure 5-8] or an SMC ladder [Figure 5-9]) to the floor tiedown (with a snaplink) inboard of the ceiling rappelling anchor point.

(3) Attach a snaplink to the ceiling rappelling ring.

(4) Route the belay rope from the backcoil, through the belay device, then up through the ceiling snaplink, and down to the cluster snaplink.

(5) Tie the end of the belay rope into a bowline with halfhitch and snap it into the cluster snaplink.

(6) The FRM identifies two ropers to push the cluster or bundle out of the aircraft. He also identifies an AFRM or a safety to belay the cluster or bundle. The FRM briefs these personnel on the time to lower the bundle, the signals to be used, and the procedure to clear the belay system before deploying the FRIES. In performing his

belaying duties, the AFRM or safety must The aviation aircrew or safety is responsible for:

- Provide sufficient slack to allow the pushers to eject the cluster from the aircraft.
- Control the descent until the cluster reaches the surface.

**Figure 5-8. Belay Device figure eight SMC** **Figure 5-9. Belay device ladder**

- When the cluster is on the surface, unsnap both the upper and lower snaplinks, drop the remainder of the rope from the aircraft, and immediately look out to make sure conditions are safe to deploy the FRIES.
- Inform the FRM and pilot when conditions are safe.

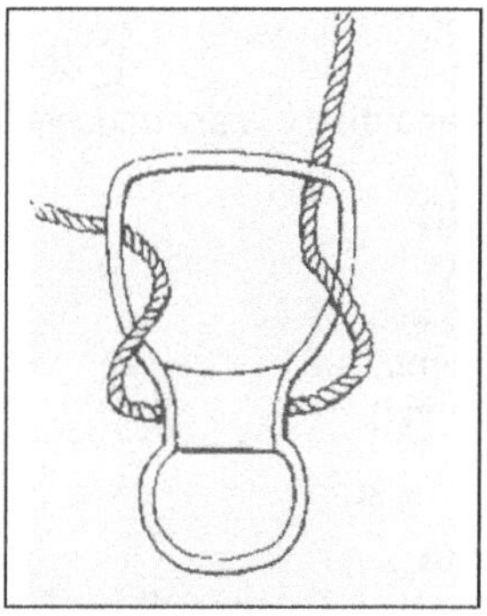

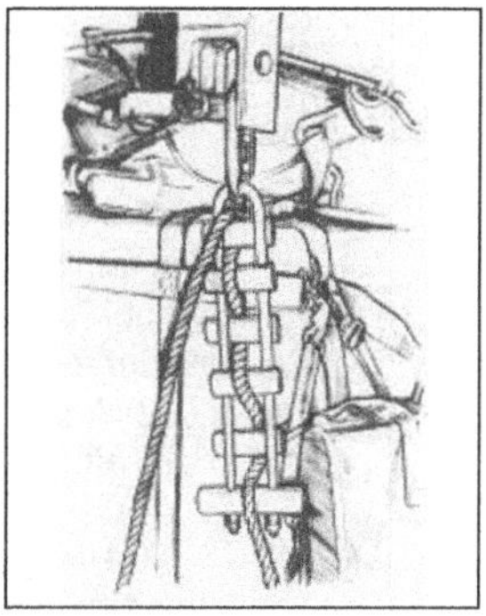

- Deploy FRIES on command

**Note:** Expect to reposition aircraft before ropers exit to avoid ropers landing on equipment.

## 5-30. HELICOPTER OPERATIONS FRM CHECKLIST

Table 5-1 shows a checklist of duties of key personnel in fast-rope operations.

| PRE-ROPING ACTIONS |
| --- |
| Receive a briefing from the OIC or the AMC. |
| Coordinate and brief all participants. |
| Rig aircraft and conduct a joint inspection. |
| Brief ropers, AFRMs, and safeties. |
| Rig and inspect ropers. |

| Conduct a static rehearsal. |
|---|
| **AIRCRAFT LOADING**<br>Position equipment and personnel.<br>Make sure all personnel have straps or seat belts.<br><br>**ACTIONS IN FLIGHT**<br>Monitor the command net.<br>Monitor the aircrew net.<br>Monitor the flight route.<br><br>**ACTIONS AT 10-MINUTE WARNING** (applies to long infiltrations)<br>Issue the 10-minute time warning and GET READY.<br>Check equipment and belay system hookup.<br>Check fast-rope hookup.<br>Secure fast-rope bar in position.<br>Make sure rope is back coiled and markers are attached.<br><br>**ACTIONS AT 6-MINUTE WARNING**<br>Issue the 6-minute time warning.<br>Remove personnel restraints or seat belts.<br>Position personnel and equipment.<br>Break chemlights (night operations).<br>Open aircraft doors, if required.<br><br>**ACTIONS AT 1-MINUTE WARNING**<br>Issue the 1-minute time warning and STAND BY.<br>Position ropers in stick formation.<br><br>**ACTIONS AT FLARE**<br>Identify the target area.<br>Deploy bundles/equipment (Safeties) and clear ropes.<br>Deploy FRIES ropes (FRM and AFRM).<br>Check ropes to make sure 15 feet of rope are on the surface (FRM). |

**Table 5-1. FRM helicopter fast-rope operations checklist**

| **ACTIONS FOR DESCENT** |
|---|
| ~ FRM, AFRM, or safety positions the number 1 man at the rope. (The FRM may exit first or last.)<br>~ FRM issues the command GO. AFRM echoes GO command, and the ropers exit the aircraft.<br>Safety informs PIC, "ROPERS OUT." |

| |
|---|
| AFRM or safety controls the rate of exit of the ropers.<br>AFRM exits last.<br>Aircrew or safeties observe the last roper. Safety tells the PIC, ALL ROPERS AWAY, after the last roper is on the surface and signals.<br>PIC issues command JETTISON or RECOVER ROPES.<br>Aircrew or safety jettisons or recovers ropes, and issues ROPES CLEAR report to the pilot. |

**Table 5-1. FRM helicopter fast-rope operations checklist (continued).**

### 5-31. FRIES COMMANDS AND SIGNALS

The following standardized arm-and-hand signals are provided for FRIES operations. Ground-to-aircraft arm-and-hand signals are military standard as found in the Pathfinder field manual 57-38.

a. Time warnings (10-, 6-, and 1-minute) are given by holding up the appropriate number of fingers and verbally sounding off.

b. The signal for GET READY is the same as the standard Airborne signal—arms extended horizontally, fingers extended vertically and joined.

c. The signal for STAND-BY is arms extended downward at a 45-degree angle from the body with hands closed except for index fingers, which are extended, pointing to the floor

d. To execute the signal for DEPLOY ROPES, the FRM or safety forms both hands into fists at mid-thigh level and sweeps arms horizontally outward. At full extension, he opens his hands as if dropping the ropes out of the helicopter and sounds off with DEPLOY ROPES.

e. The signal for GO is arms raised to horizontal, hands closed with index fingers extended pointing to the ropes.

f. The signal for STOP or ABORT is hands closed into fists with arms raised across the forehead .

g. The signal for SIT-DOWN and DON SEAT BELTS is pointing to the floor, followed by moving both fists around waist, as if donning a belt.

h. The signal for OK is hands raised to eye level, index finger and thumb forming a circle with other fingers extended.

## Section IX. EMERGENCY ACTIONS

Procedures for correcting an emergency condition that could reasonably be encountered are described in this section. Multiple emergencies, adverse weather, or other unusual conditions may require modification of these procedures. The nature and severity of the emergency dictate the response necessary; therefore, personnel must use sound judgment in determining the correct action to be taken.

## 5-32. EMERGENCIES BEFORE ROPING STARTS

All personnel sit down, don their seat belts, and take further instructions from the pilot or crew chief.

## 5-33. EMERGENCIES AFTER ROPING STARTS

Emergency procedures to be followed after roping starts are discussed in this paragraph.

a. **Unsafe Drift or Premature Liftoff.** Anyone who detects the aircraft having drifted off the site must immediately stop training and inform the FRM, AFRM , safety, or PIC.

(1) The FRM, AFRM, safety, or roper stops stick.

(2) The ropers stop descent and lock in.

(3) The FRM or crewman informs the PIC, and guides him in moving the aircraft back on target.

(4) The unit continues operations.

b. **Rope Hung or Snagged (On Ground, Tree, Building).** Anyone who detects the rope being snagged or hung must immediately stop training and inform the FRM, AFRM, safety, crew chief, or PIC.

(1) The safety ensures ropers are off the rope and are clear.

(2) The aircraft descends or lands as needed.

(3) The safety frees the rope and informs the pilot.

(4) The unit resumes the operation.

c. **Lost Communications.** All training and operations must be planned to include the use of the intercom between the PIC or crewmen and the FRM. If the intercom fails, the following arm-and-hand signals can be used until the rope can be cleared and the intercom is restored.

- Wrists crossed diagonally above the forehead with clenched fists to signal stop stick.
- Finger pointed toward the exit and a verbal GO to signal ropers to exit.
- Open palm moved and faced in the required direction to signal aircraft movement.
- Clenched fist to stop aircraft movement.

**Note:** Training will not be conducted without radio communications between the pilots or aircrew and the FRM and the ground.

# Section X. SAFETY

Safety is everyone's responsibility. All personnel involved in FRIES operations are responsible for identifying hazardous situations and preventing injuries to personnel. Anyone who observes an unsafe condition or act is authorized to halt the operation and inform the FRM or the PIC.

## 5-34. PREFLIGHT AND INFLIGHT

Anyone who detects an inflight emergency will immediately notify the PIC, crew chief, or safety and follow aircrew instructions.

a. Before a FRIES operation, the FRM will coordinate with the PIC both routine and emergency procedures as well as any unique conditions or potential problems. He also discusses the procedures to be used during flight, during roping, and before the aircraft's departure from the roping site. The FRM then discusses the information with all the AFRMs, safeties, and crew chiefs as well. He reviews the time warnings (including arm-and-hand signals) and appropriate actions such as who will deploy ropes, verify rope status, and perform rope-clearing procedures.

b. The mission prebrief covers any aircraft equipment that, if disturbed, could cause an aircraft or fast-rope problem. The problem areas should be made known to all personnel concerned and preventive measures taken to neutralize or eliminate them where possible.

c. The FRM gives the crew chief the final position adjustment over the target, using hand signals.

d. All safety straps and communications cords that are not in use are securely stowed. All safety straps and communication cords in use should be positioned so they do not present a hazard to personnel movement in the aircraft.

e. After the PIC gives the command DEPLOY ROPES, the FRM is responsible for deploying the rope and making sure it is in contact with the surface before ropers exit.

**Note:** The safety is responsible for observing rope and ropers and for making sure the pilot maintains aircraft position until all ropers are on the surface.

**CAUTION**

**According to AR 95-3, anytime operations occur over water, all personnel will wear an approved personal flotation device as identified in FM 57-220 or USASOC Regulation 350-2.**

## 5-35. DURING ROPING

The following safety procedures apply during roping.

a. During the actual descent, fast-ropers will not use NVGs because they limit depth perception and create a tunnel vision effect.

b. During descent, ropers must watch for obstructions and lower ropers, and must break as necessary.

c. Individual ropers will lock in during emergencies by wrapping one leg around the rope once or twice and standing on the fast rope with the other foot.

d. Ropers must maintain sufficient intervals on the rope to prevent contact with lower ropers. Individual ropers exit when the head of the man in front of them disappears below the ramp.

e. Ropers execute descents at speeds commensurate with their experience and proficiency in FRIES infiltration operations. (For example, less-experienced ropers should descend slowly until they become more proficient.)

f. The number of ropers on each rope at one time must not exceed the safe limit of the rope mount point in use, as follows:

(1) Do not exceed 1,500 pounds per rope on the MH/UH-60.
(2) Do not exceed 1,500 pounds per rope on the MH/UH-47.
(3) Do not exceed 1,500 pounds per rope on the MH/UH-53.
(4) Do not exceed 1,300 pounds per rope on the MH-60G.

## 5-36. SAFETY BRIEFING

All units must present a safety briefing before conducting FRIES training or operations. The safety briefing must cover the following (as a minimum):

- The planned event (day or night, tower or helicopter, high or low hover, insertion or extraction).
- Actions on the aircraft (or tower).
- Operation (or training) location and any hazards within 500 meters.
- Identification of key personnel, their location, and their duties.
- FRIES equipment and its characteristics.
- FRIES equipment rigging or inspection.
- Roping personnel preparation (shirttails tucked in, chin strap fastened, loose straps taped, gloves and goggles on).
- Arm-and-hand signals.
- Emergencies and emergency procedures.
- Medical procedures (medic with aid bags, backboards, litters, and dedicated vehicle with driver on site).
- Communications requirements (frequencies, call signs, required reports).
- Any other information essential to safety.

## 5-37. SAFETY REMINDERS

The following safety reminders will be reviewed by all key personnel before conducting operations.

a. Ropers must wear long-sleeve shirts or jackets, full-length pants, laced boots, helmets, protective goggles, and heavy leather gloves.

b. MILES harnesses must not be worn during fast-roping.

c. Ropers will not wear NVGs.

d. Commanders must personally approve the rope site selected for night operations. The FRIES site must be large enough to permit all ropers, upon reaching the surface, to move clear of the rope and for units to consolidate or re-form.

e. All night rope operations must be assessed as medium risk or higher.

f. FRIES descent training is not authorized with a fast rope longer than 90 feet.

g. Ropers must not carry more than 50 pounds of equipment during FRIES descents. The maximum weight of rucksacks is limited to 35 pounds. When rucksacks exceed 35 pounds (or total load exceeds 50 pounds), gear must be lowered (belayed) on a separate system, preferably from the opposite door.

h. Ropers will not carry bulky equipment that could interfere with FRIES operations. Equipment that cannot be carried by the soldier during roping (due to its weight or bulk) can be lowered (belayed) during insertion. During extractions, rucksacks must be attached to the extraction loops of the FRIES rope using snaplinks.

i. Ropers must not make more than 6 combat equipment descents or a total of 10 descents in a 24-hour period.

j. FRIES training must not be conducted if—

- The aircraft is unstable.
- The fast rope is not fully deployed (15 feet of the rope must be on the surface).
- Any obstacle is present that could interfere with the descent of the ropers.
- The rope is frozen or slippery
- Anyone identifies any unsafe condition.

k. Safety personnel must—

- Position themselves so they can observe active FRIES points with out obstructing ropers.
- Conduct final rigging checks of the mount, equipment, and ropes at the 10-minute warning.
- Ensure the aircraft is over the target area and the ropes reach the ground.
- Constantly keep both the PIC and FRM informed—
  - When the aircraft is in position for deployment.
  - When anything is being deployed from the helicopter (ropes, bundles or personnel).
  - When 15 feet of rope are on the surface.
  - When the first roper exits on the fast rope.
  - When all ropers on the surface are clear.
  - When the ropes are recovered or jettisoned.
  - When the aircraft is cleared for flight.
  - When 15 feet of the extraction rope is on the surface.
  - When extraction personnel are attached to the rope and ready for lift off.
  - When extracted personnel are safely lowered to the ground.
  - When extracted personnel are out from under the aircraft.
  - Any other conditions that could impact safe operations.

**Note:** If the FRM and the safety exit, the crew chief performs rope clearing and or jettisoning duties.

**WARNING**

**Using the word CLEAR tells pilots the aircraft is ready to fly. Before any use of the word CLEAR, FRMs and safeties must ensure that all personnel are well away from the ropes and the ropes have been fully recovered or released (during insertion), or all personnel are securely attached to the ropes (during extraction).**

l. FRIES training must not be conducted in densely wooded areas or areas prone to blowing dust, sand, or snow, which can produce a hazard for ropers or the aircraft.

m. FRIES training and operations must not be conducted during severe weather such as rain, sleet, ice, or extreme cold.

n. The possibility of personnel injury during FRIES training and operations is always present. Every soldier involved in FRIES activities must maintain constant vigilance to prevent personnel injury and equipment damage.

CHAPTER 6

# EQUIPMENT

*This chapter discusses the basic equipment used for rappelling operations (Appendix B). It covers the types of ropes, the procedures for rope selection, and the care for each rope type. It also discusses the two methods of coiling and the different snaplinks used to join the equipment and rope.*

## Section I. ROPES

Ropes are the most important items of equipment for the rappeller. They provide access down the obstacle or aerial platform, while ensuring individual safety.

### 6-1. TYPES OF ROPES

Two types of ropes are available to the rappeller—*static* and *dynamic*.

a. **Static Rope.** This rope allows for minimal stretch of the rope. The static rope stretches about 5 to 15 percent at the point of failure and about 2 percent under a working load. The minimum tensile strength for an 11-mm (7/16-inch) static rope for military use is 4,500 pounds. *Kernmantle* static ropes range from 3 millimeters to 11 millimeters in diameter and are constructed similar to the 550-parachute cord. (*Kern* means core and *mantle* means sheath.) The internal core of the rope is constructed of a continuous multifilament-nylon yarn, which is spun into continuous parallel strands. This internal core is then covered with a nylon-braided outer sheath. Due to the internal parallel strand construction of the static rope, there is less spinning and kinking than with a dynamic rope. Static ropes can be used for slings, harnesses, etriers (stirrups), rappelling, rope installations, hauling lines, and where stretch of the rope is not desired.

b. **Dynamic Rope.** This rope allows for stretch within the fibers of the rope, which can be a disadvantage in rappelling, prusik climbing, and other applications. Dynamic ropes are more susceptible to abrasion and wear. They have about a 5 to 10 percent working stretch. The minimum tensile strength for an 11-mm (7/16-inch) rope for military use is 4,500 pounds. Two types of dynamic ropes are nylon-laid and kernmantle:

(1) *Nylon-laid ropes.* Synthetic fibers are the best climbing and rappelling material. Nylon has become the standard material for climbing ropes and has replaced manila, flax, hemp, and sisal. The military mountaineering critical application rope is constructed in a *laid* fashion with continuous multifilament-nylon fiber yarn twisted into strands. Three strands are twisted into a climbing rope of a specific diameter ( Figure 6-1). Nylon-laid ropes are easy to inspect for serviceability by twisting the fibers. Laid ropes tend to untwist slightly when under a load, causing kinking and spinning. They are also susceptible to abrasion. The following specifications pertain to the standard military nylon-laid climbing/rappelling rope:

- 36 1/2 meters (120 feet) long.
- 11 millimeters (7/16 inch) wide.
- At least a 4,500-pound tensile strength.
- Right-hand lay.
- One-third stretch factor (at point of failure).

- May lose as much as 15 percent of its strength when wet.
- Weighs 6 pounds when dry.

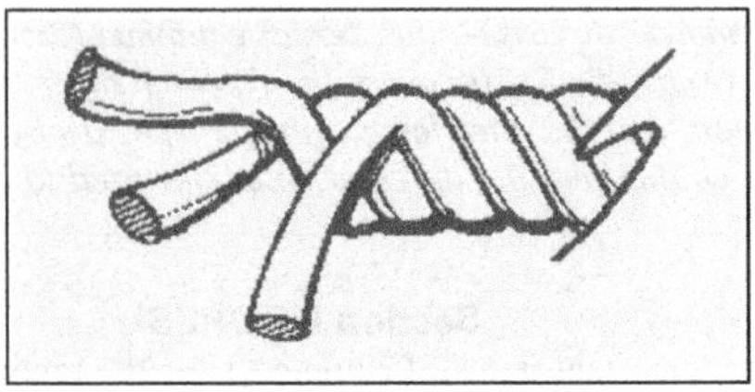

**Figure 6-1. Nylon-laid rope.**

(2) *Kernmantle ropes.* These ropes are similar to their static counterparts in that they consist of an inner core and an outer sheath. The core is constructed of continuous twisted nylon filaments, which are laid or braided together and enclosed in a tightly braided outer sheath. These ropes are well suited for climbing on rock, snow, or ice where a brief elastic stretch of the rope occurs during a fall. They have a stretch of about 40 percent at the point of failure. The breaking strength is high, and there are no exposed strands for rock crystals to work between to damage the rope ( Figure 6-2). There is less sliding friction through the snaplink and over other surfaces, since the outer sheath is smooth. Kernmantle ropes are available in many sizes, lengths, stretch factors, tensile strengths, and fall ratings.

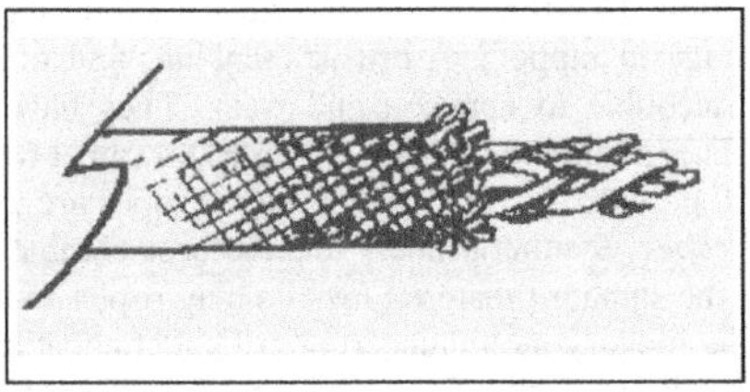

**Figure 6-2. Kernmantle (core and sheath) rope.**

**6-2. ROPE SELECTION**

Rope selection for military operations requires ropes of various types, lengths, and diameters; and depends on its intended use (static or dynamic), environment (water-repellant), climate, mission (lightweight ropes), and other factors. Static ropes should have minimal stretch; be smooth, flexible, and strong; resist abrasion and cutting; and have a low-impact force.

a. *Laid rope* is easier to inspect (by twisting the strands apart to observe the interior surfaces), has greater friction in holding knots, and is less expensive than kernmantle ropes. Its disadvantages include: much lower abrasion resistance, extremely high stretch factor for both static and dynamic loads, greater friction on rock, readiness to kink, tendency to spin a free-hanging rappeller, and greater absorption of water and dirt.

b. *Kernmantle* rope is easy to work with, has less friction, does not kink easily, and has less susceptibility to abrasion. The outer sheath protects the internal fibers from dirt, abrasion, and ultraviolet radiation. Its disadvantages are that it is hard to determine damage to the core fibers, and it is more expensive than laid rope. Kernmantle rope is not readily identifiable as dynamic or static and must be marked accordingly.

c. Rope diameter and length vary with intended use. A standard diameter for many uses is 11 millimeters (7/16 inch). For stirrups, utility ropes, and hauling lines, 7 millimeters (5/16 inch), which can also be used in double strands, is adequate. Standards to remember are that the smaller the diameter, the less the tensile strength; but the larger the diameter, the heavier the rope. The standard lengths are 36 1/2 meters (120 feet), 40 meters (135 feet), 45 meters (150 feet), and 50 meters (165 feet). Normally, standard lengths should not be cut. If a nonstandard length is needed, the rope should be precut before the mission and clearly marked.

d. A rope should be selected for a specific use. There are different types of ropes for various uses. A dynamic rope is designed for climbing, while a static rope is designed for rappelling, rescue operations, load hauling, and rope installations.

**WARNING**

**Using a rope for other than its intended use could result in personal injury or damage to equipment.**

## 6-3. CARE OF ROPES

Because the rope is the rappeller's lifeline, it requires a great amount of care and maintenance.

a. Before and after use, the rope should be carefully inspected for excessive wear, cuts, mildew, or rotten spots. If any of these defects are found, the rope is unserviceable. An unserviceable rope is destroyed by cutting it into usable parts. If no sections are usable, the rope is cut into smaller parts and discarded.

b. Ropes should never be spliced together, stepped on, or dragged. This greatly reduces the tensile strength of the rope.

c. The rope must not come in contact with sharp edges. If this is likely, the areas must be padded or taped before rappelling.

d. The rope must be kept as dry as possible. If it becomes wet, it must be dried out to prevent rotting. (Avoid excessive heat/heating during the drying process.)

## 6-4. COILING

Two methods of coiling are the *mountaineer's coil* and the *farmer's coil* ( Figure 6-3).

a. **Mountaineer's Coil.**

(1) To use the mountaineer's coil, grasp the rope about 3 feet from the end with the left hand. The right hand is then placed next to the left hand and runs along the rope until both arms are outstretched. Grasping the rope firmly, bring the hands together to form a loop, which is laid in the left hand. Repeat, forming uniform loops that run in a clockwise direction until the rope is completely coiled. If the rope tends to twist or form figure eights, give it a slight twist with the right hand when each loop is formed. Coil the rope in a clockwise direction to conform with the lay.

(2) To complete the coil, form a bight about 30 centimeters (12 inches) long with the starting end of the rope and lay it along the top of the coil. Uncoil the last loop and use it to wrap around the coil and the bight. Wrap toward the closed end of the bight, binding the final wrap across itself to lock it in place. Make six to eight wraps to secure the coil. Push the end of the rope down through the closed end of the bight. Carry the running end of the coil either in the pack (by forming a figure eight, doubling it, and placing it under the flap) or by placing it over the shoulder and under the opposite arm. If the rope to be coiled is anchored as in coiling a belay or rappel lane, start the coil near the end closest to the anchor so that the kinks work themselves out of the free end.

b. **Farmer's Coil.**

(1) The farmer's coil is easier and faster to coil than the mountaineer's coil. Establish the center of the rope either by locating its center mark or by grasping both ends of the rope and feeding them out until a bight comes up that is the center of the rope. Begin coiling with the doubled rope the same as the mountaineer's coil method. Leave about 4.6 meters (15 feet) of rope uncoiled. Then, squeeze the coils together and make four to six wraps around the middle of the coil, ensuring that the first wrap is held in place by the other wraps.

(2) Form a bight with the two running ends and place it through the bight formed by the top of the coil. Then, run the two running ends over the top of the coil and through the bight that is formed. Dress them in the center of the back of the carrier. Run the two ends over the shoulders to form shoulder straps. Then, bring them under the arms, cross them in the back over the coil, bring them around the body of the carrier, and tie them off with a square knot at the stomach.

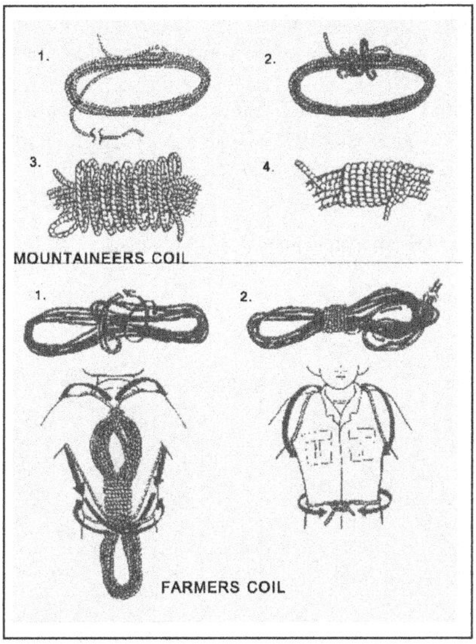


**Figure 6-3. Two methods of tying a coil.**

## Section II. SNAPLINKS

Snaplinks are used to join equipment, rope, and people into a functioning system. When used properly, they are strong, versatile items. Snaplinks are metal (steel, aluminum, or alloys) loops with a hinged, spring-loaded gate on one side. They come in many sizes, shapes, and strengths (with and without locking gates) and in many types of metal. Steel is the strongest snaplink, but it is heavier.

### 6-5. DESCRIPTION

Snaplinks are available in many sizes and shapes to suit different needs. A locking snaplink is best for rappelling. Nonlocking snaplinks are easy to operate, but should be used only where they cannot be accidentally opened. Hollow snaplinks should be avoided since their use is limited. The following information applies to all snaplinks.

a. The weakest part of a snaplink is the gate. The gate must be closed before applying a load. When the gate is open, snaplinks should have little or no lateral movement of the gate.

b. Locking pins should be checked to ensure that they are not loose, worn, or corroded.

c. The metal should be checked for any cracks, grooves, burrs, flaws, or rust.

d. The spring-loaded gate should automatically close securely from an open to a closed position with no gap between the locking pin and notch.

e. If an engraver is used to mark snaplinks, it should be applied only to the gate, never to the load-bearing side.

f. A snaplink should never be side-loaded (across the gate) since this reduces the overall strength to the point of gate failure.

## 6-6. TYPES OF SNAPLINKS

The two types of snaplinks used during military operations are the standard snaplink (Figure 6-4) and the locking gate snaplink ( Figure 6-5). Both snaplinks are available in a D-shaped variation or an oval variation.

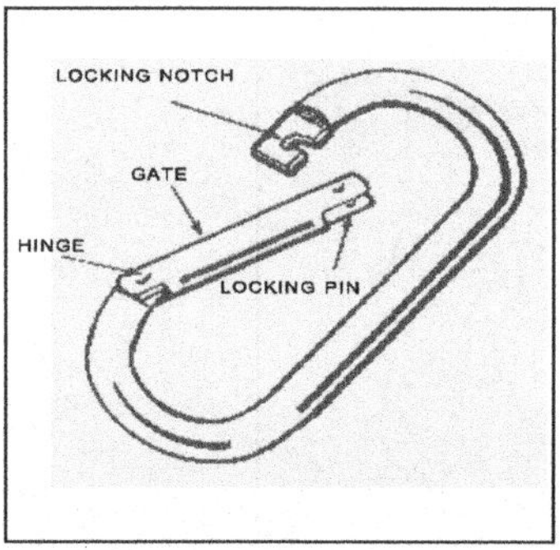


**Figure 6-4. Standard snaplink (oval).**

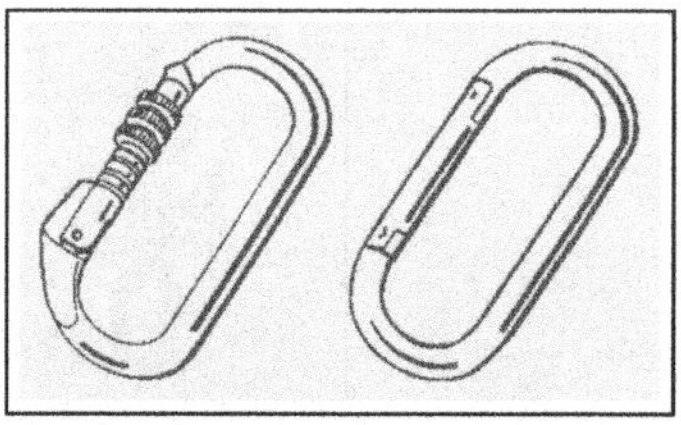

**Figure 6-5. D-shaped snaplinks.**

a. **D-Shaped Snaplinks.** D-shaped snaplinks are stronger than the oval type because the shape directs the largest part of the load to be applied to the longer, stronger side opposite the gate (Figure 6-5). D-shaped snaplinks are made of steel or aluminum alloys, and are available in many sizes and thicknesses, all of which can be with or without locking gates. Their strength and durability vary, which must be considered before use.

b. **Oval Snaplinks.** Oval snaplinks are versatile and have different applications. They are made of steel or aluminum alloys and are available in many sizes and thicknesses. Both sides of an oval snaplink bear the strain equally under load weight. Many modified ovals are available and most can be with or without the locking gate. All must meet military standards or UIAA specifications.

c. **Locking Snaplinks.** Locking snaplinks have a locking mechanism with a threaded sleeve on the gate. The sleeve screws tightly over the gate opening end or hinge end to hold the gate closed (Figure 6-6). A reverse locking gate is needed to prevent a moving rope from unscrewing the sleeve. The locking sleeve and threads should be kept free of dirt and grit. If the sleeve is forced to close, it may strip the threads. The locking mechanism ages and weakens after repeated use and should be routinely inspected for wear. The characteristics of locking snaplinks are as follows:

- Material: steel.
- Approximate strength: 2,000 pounds with the gate closed.
- Gate: spring-loaded without locking sleeve.
- Weight: 4.3 ounces.

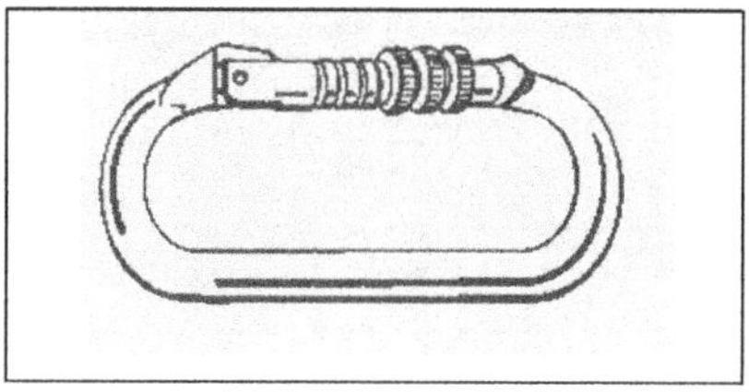

**Figure 6-6. Locking snaplink.**

## 6-7. INSPECTION

All rappelling material and individual rappelling equipment must be inspected before conducting a rappelling operation. The rappel master is responsible for ensuring the serviceability of rappelling material and individual rappelling equipment.

a. **Snaplinks.** Snaplinks should be inspected daily before, during, and after use.

(1) The metal should be checked for cracks, grooves, burrs, rust, and flaws. The gate should open and close freely without binding. There should be no lateral movement when the gate is open. The gate spring action should snap shut when released. The locking notch should have a slant or slot so that the gate remains shut under the impact of a rappeller's fall. The gate pins should not work their way out of their holes and should not be shorter than their holes. If there is a locking mechanism, it should be inspected to ensure that threads are not stripped and that the sleeve tightly locks the gate.

(2) If burrs, grooves, or rough areas are identified, the snaplink should not be used. Rust should be removed with steel wool and the spot rubbed with oil or solvent. The spring should be lubricated as needed. The snaplink must be boiled in water for 20 to 30 seconds to remove the cleaning agents since solvents and oils cause dirt to cling to the snaplink and to rub off on the ropes. It is better to use a dry graphite-based lubricant on snaplinks since such lubricants do not attract dirt.

b. **Ropes.** The ropes are inspected for serviceability. A rappel rope is unserviceable if it is saturated with petroleum products, is mildewed, shows excessive wear, has soft spots (internal damage), or if one strand is cut more than one-half of its diameter. Ropes should be inspected daily—before, during, and after use. An applicable annotation should be made in a corresponding rope log (DA Form 5752-R) ( Figure 6-7).

| ROPE LOG (USAGE AND HISTORY)<br>For use of this form, see TC 90-6-1, the proponent agency is TRADOC | | | UNIT ID MARKING<br>D-11-86-5 |
|---|---|---|---|
| NSN<br>8455-73560366 | DOCUMENT NUMBER<br>W330SV 61890007 | SERIAL NUMBER<br>6-6-6933 | MFR LOT NUMBER<br>G620135115103 |
| DATE OF MFR<br>08 Feb 87 | ISSUE DATE<br>01 OCt 87 | DATE IN SERVICE<br>17 NOV 87 | LENGTH<br>45M (150 FT) |
| DIAMETER<br>11mm (7/16 inch) | FIBER<br>TYPE 6.6 NYLON | COLOR<br>OD Green | CONSTRUCTION<br>DYNAMIC KERMANTLE |

INSPECT ROBE FOR DAMAGE OR EXCESSIVE WEAR EACH TIME IT IS DEPLOYED AND AGAIN AFTER EACH USE. IMMEDIATELY RETIRE ALL SUSPECT ROPES.

| DATE USED | LOCATION | TYPE OF USE | ROPE EXPOSURE | INSPECTOR'S INITIAL/DATE | ROPE CONDITION AND COMMENTS |
|---|---|---|---|---|---|
| 20 Nov 87 | Yonah Mtn GA | Direct Aid Climb 5.7 A1 | 65°F clear & dry Smoothe Granite | DLE 20 Nov 87 | |
| 21 Nov 87 | Yonah Mtn GA | Top Rope Belayed Balance Climb | 68°F Clear & dry Smoothe Granite | DCY 21 Nov 87 | |
| 15-18 Jan 88 | Gulcana Glacier ALASKA | ROPED Glacier Travel | 2°F snow storm | JEY 15-18 Jan 88 | |
| 27 Jan 88 | Smugglers Notch VT | Front Point Ice Climb | 35°F Cloudy No Precip wet rope Due to ice than | ATS 27 Jan 88 | SUSTAINED ONE 15 FT Leader FALL |
| 28 Jan 88 | Smugglers Notch VT | Front Point Ice Climb | 20°F Cloudy & Snowy | JML 28 Jan 88 | |
| 8 Feb 88 | Mt Shasta CAL | Flat Foot Ice Climb | 8°F Lightly Snowing | DCY 8 Feb 88 | Approx one inch of Sheath Slippage on End of rope |
| 22-24 Feb 88 | DECOS MTN NM | TWO MAN PARTY Rock Climb 5.7+ | 45°F Clear & DRY GRANITE | [illegible] 22 Feb 88 | |
| 10-22 Jan 88 | Mt McKinley ALASKA | ROPED PARTY Climb CLASS VI | 20-53°F ONE DAY OF RAIN | JMK 10-22 Jan 88 | Abrasion to Rope at points of TIE-IN (Ends & middle) |

DA FORM 5752-R, MAY 89

**Figure 6-7. Example of completed DA Form 5752-R.**

c. **Gloves.** The gloves are inspected for serviceability. They are unserviceable if holes are found in the friction-bearing surfaces or along the seams.

d. **Helmet.** The helmet is inspected for correct assembly and serviceability. It is unserviceable if it cannot be properly assembled or if any straps are cut or torn one-quarter of its width.

(1) Ensure that the chin strap attaching tabs are secured to the helmet by the locking nuts.

(2) Ensure that the pull-the-dot tab has four plies of nylon and that the chin strap is routed from the inside of the buckle around the horizontal bar and back toward the inside of the buckle.

## Section III. ALTERNATE METHODS OF DESCENT

Snaplinks or carabiners that are used for rappelling are placed through the sewn loops of the seat harness or the sewn loops of the seat-chest combination.

### 6-8. FIGURE-EIGHT DESCENDER

The rappeller attaches a locking carabiner to the harness (Figure 6-8). He routes the rappel rope up through the large hole and places the bight over the collar. He inserts the figure eight with the rappel rope attached into the locking carabiner, and locks it down. The rappeller brakes to the rear and descends as in a seat hip rappel. If heavy loads are to be descended, he uses a double wrap around the collar.

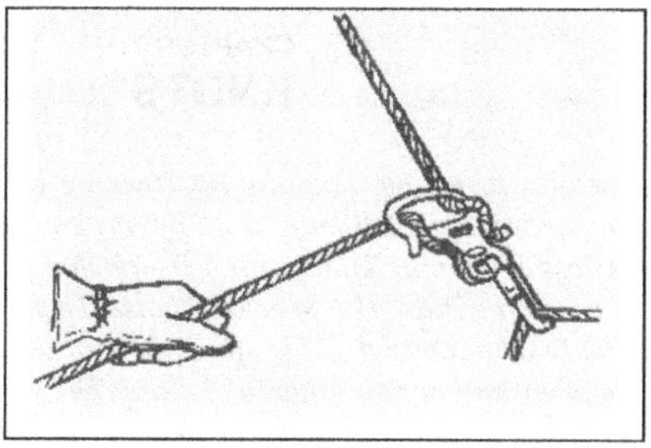

Figure 6-8. Figure-eight descender.

## 6-9. MUNTER HITCH

The rappeller attaches a large radius snaplink to the harness or rappel seat. He ties a Munter hitch and clip into the snaplink (Figure 6-9, page 6-9). He brakes to the front and descends with an L-shape body position.

**Notes:** 1. The guide hand should remain on the standing end of the rope to keep the rappeller's head and face away from the hardware.

2. The Munter hitch creates significant rope-to-rope friction and may cause premature wear on nylon ropes.

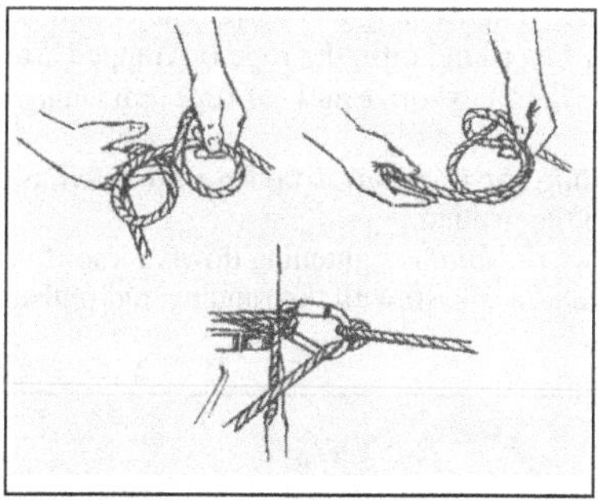

Figure 6-9. Munter hitch.

CHAPTER 7

# KNOTS

*A knot is a fastening made by intertwining or tying pieces of string, cord, rope, or webbing. All knots used by a rappeller are divided into four classes: Class I—joining knots, Class II—anchor knots, Class III—middle rope knots, and Class IV—special knots. These classes of knots are intended only as a general guide since some of the knots discussed herein may be appropriate in more than one class. The skill of knot tying may be lost if it is not used and practiced. With experience and practice, knot tying becomes instinctive and helps the soldier in many situations.*

## 7-1. MILITARY MOUNTAINEERING TERMINOLOGY

The terms explained below are the most commonly used in military mountaineering (Figure 7-1).

a. A *bight of rope* is a simple bend of rope in which the rope does not cross itself.

b. A *loop* is a bend of rope in which the rope does cross itself.

c. A *half hitch* is a loop that runs around an object and locks itself.

d. The *running end* (working end) of the rope is the free end of the rope that can be used.

e. The *standing part* of the rope is the part that is static (anchored, coiled); the remaining part of the rope not being used (also called static end).

f. The *lay* of the rope is the same as the twist of the rope.

g. A *round turn* is a single complete wrap of the rope around an object providing 360-degree contact. The running end leaves the completed circle in the same direction as the standing part. In a round turn, the rope is wrapped around an object 1 1/2 times.

h. The *pigtail* is the short length of rope remaining at the end after tying a knot or coiling a rope.

i. *Back feeding* (or stacking) is taking off one wrap at a time from a coil, and letting it fall naturally to the ground.

j. *Dress down the knot* is tightening down a knot to its functioning form.

k. An *overhand* is a loop with the running end pulled through the loop.

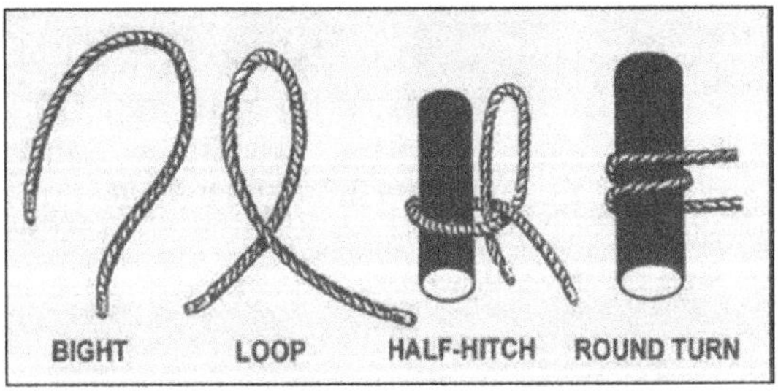


**Figure 7-1. Examples of rope terminology.**

## Section I. JOINING KNOTS

### 7-2. SQUARE KNOT

The square knot is used to tie the ends of two ropes of equal diameter (Figure 7-2).

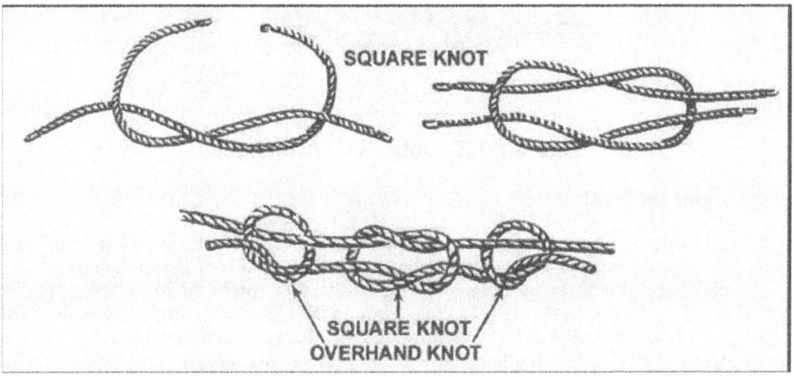


**Figure 7-2. Square knot.**

a. **Tying the Knot.**

STEP 1. Holding one working end in each hand, place the working end in the right hand over the one in the left hand.
STEP 2. Pull it under and back over the top of the rope in the left hand.
STEP 3. Place the working end in the left hand over the one in the right hand and repeat STEP 2.
STEP 4. Dress the knot down and secure it with an overhand knot on each side of the square knot.

b. **Checkpoints.**

(1) There are two interlocking bights.

(2) The standing parts are on the same side and properly secured with overhand knots.

### 7-3. DOUBLE SHEET BEND KNOT

The double sheet bend knot is used to tie the ends of two or more ropes of equal or unequal diameter (Figure 7-3). When a single rope is tied to a multiple of ropes, the bight is formed with the multiple of ropes.

Figure 7-3. Double sheet bend knot.

a. **Tying the Knot.**

STEP 1. Form a 30-centimeter (12-inch) bight in the left hand with the rope or ropes. Ensure that the short end of the bight is facing to the inside.
STEP 2. Place the index finger of the left hand on top of the bight. Bring the rope in the right hand up through the bight and over the index finger of the left hand, so the working end is away from the body.
STEP 3. Reach through the loop and grasp the working end of the rope and bring it back towards the body. Place it between the index finger and the bight, forming a round turn. Repeat this one more time, wrapping toward the tip of the index finger.
STEP 4. Remove the finger and dress the knot down.

b. **Checkpoints.**

(1) The two wraps around the bight are held in place by a locking bar.

(2) The two standing parts of the ropes exit parallel and from opposite ends of the knot. The two working ends form an L and exit the knot at a 90-degree angle and on opposite sides of the knot.

## 7-4. FISHERMAN'S KNOT

The fisherman's knot is used to tie two ropes of similar or dissimilar materials (Figure 7-4).

Figure 7-4. Fisherman's knot.

a. **Tying the Knot.**

STEP 1. Tie an overhand knot in one end of the rope.
STEP 2. Pass the working end of the other rope through the first overhand knot. Tie an overhand knot around the standing part of the first rope with the working end of the second rope.
STEP 3. Tightly dress down each overhand knot and tightly draw the knots together.

b. **Checkpoints.**

(1) The two separate overhand knots are tied tightly around the long standing part of the opposing rope.

(2) The two overhand knots are drawn snug.

## 7-5. DOUBLE FISHERMAN'S KNOT

The double fisherman's knot (also called double English or grapevine) is used to tie two ropes of similar or dissimilar materials (Figure 7-5).

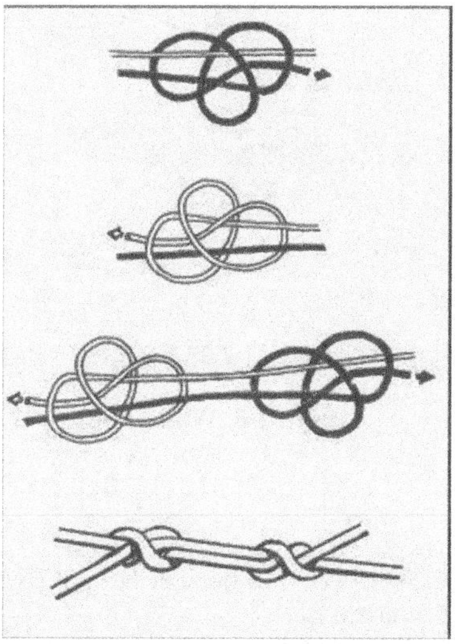

**Figure 7-5. Double fisherman's knot.**

a. **Tying the Knot.**

STEP 1. With the working end of one rope, tie two wraps around the standing part of another rope.
STEP 2. Insert the working end (STEP 1) back through the two wraps and draw it tight.

STEP 3. With the working end of the other rope, which contains the standing part (STEPS 1 and 2), tie two wraps around the standing part of the other rope (the working end in STEP 1). Insert the working end back through the two wraps and draw tight.

STEP 4. Pull on the opposing ends to bring the two knots together.

b. **Checkpoints.**

(1) Two wraps with the working end running through and drawn tight.

(2) Both wraps are drawn together with all the slack out of the knot.

## 7-6. WATER KNOT

The water knot is used to attach two webbing ends (Figure 7-6). It is also called a ring bend, overhand retrace, or tape knot. It is used in runners and harnesses.

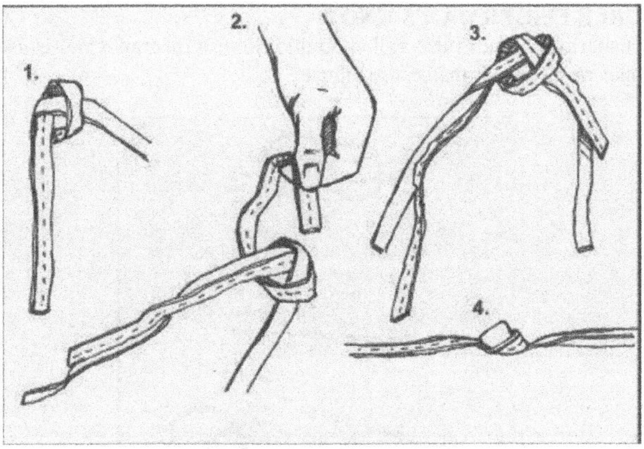


**Figure 7-6. Water knot**

a. **Tying the Knot.**

STEP 1. Tie an overhand knot in one of the ends.

STEP 2. Feed the other end back through the knot, following the path of the first rope in reverse.

b. **Checkpoints.**

(1) There are two overhand knots, one retracing the other.

(2) There is no slack in the knot, with the working ends coming out of the knot in opposite directions (at least 5 centimeters [2 inches] long).

## Section II. ANCHOR KNOTS

### 7-7. BOWLINE KNOT

The bowline knot is used to tie a single fixed loop in the end of a rope (Figure 7-7). It does not slip under strain and is easily untied. This knot is always used when there is alternating tension. It can also be used to tie the end of a rope to an anchor.

**Figure 7-7. Bowline knot.**

a. **Tying the Knot.**

STEP 1. Bring the working end of the rope around the anchor, from right to left (as the rappeller faces the anchor).

STEP 2. Form an overhand loop in the standing part of the rope (on the rappeller's right) toward the anchor.

STEP 3. Reach through the loop and pull up a bight.

STEP 4. Place the working end of the rope (on the rappeller's left) through the bight, and bring it back onto itself. Dress the knot down.

STEP 5. Form an overhand knot with the tail from the bight.

b. **Checkpoints.**

(1) The loop is locked into place by a bight.

(2) The short portion of the bight is on the inside and on the loop around the anchor (or inside the fixed loop).

**7-8. ROUND TURN AND TWO HALF HITCHES KNOT**The round turn and two half hitches knot is used to tie the end of a rope to an anchor and must have constant tension (Figure 7-8).

**Figure 7-8. Round turn and two half hitches knot.**

a. **Tying the Knot.**

STEP 1. Route the rope around the anchor from right to left and wrap down (must have two wraps in the rear of the anchor and one in the front). Run the loop around the object to provide 360-degree contact, distributing the load over the anchor.

STEP 2. Bring the working end of the rope left to right and over the standing part, forming a half hitch (first half hitch).

STEP 3. Repeat STEP 2 (last half hitch has a 15-centimeter [6-inch] pigtail).

STEP 4. Dress the knot down.

b. **Checkpoints.**

(1) A complete round turn should exist around the anchor with no crosses.

(2) Two half hitches should be held in place by a locking bar with no less than a 15-centimeter (6-inch) tail remaining.

(3) When the two half hitches are twisted to the left or the right, they will appear to for a figure eight.

(4) Looking at the top of the half hitches, the rope should appear to be one continuous rope.

**7-9. FIGURE-EIGHT RETRACE KNOT**

The figure-eight retrace knot is also called a rerouted figure eight. Although it produces the same result as a figure-eight loop, by tying the knot in a retrace it can be used to fasten the rope to trees or to places where the loop cannot be used (Figure 7-9).

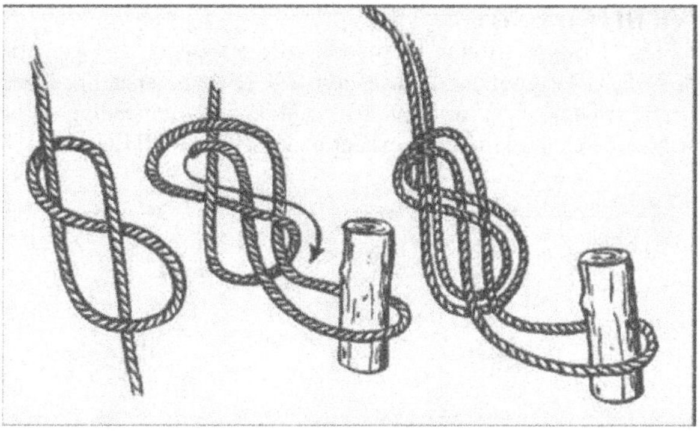

**Figure 7-9. Figure-eight retrace.**

a. **Tying the Knot.**

STEP 1. Use a length of rope long enough to go around the anchor, leaving enough rope to work with.

STEP 2. Tie a figure-eight knot in the standing part of the rope, leaving enough rope to go around the anchor. To tie a figure-eight knot:

- Form a loop in the rope.
- Wrap the working end around the standing part.
- Route the working end through the loop.
- The finished knot is dressed loosely.

STEP 3. Place the working end around the anchor point.

STEP 4. With the working end, insert the rope back through the loop of the knot in reverse.

STEP 5. Keep the original figure eight as the outside rope and retrace the knot around the wrap and back to the long sttanding part.

STEP 6. Safety off with an overhand knot.

b. **Checkpoints.**

(1) At least 15 centimeters (6 inches) of rope extends past the figure-eight knot.

(2) The original figure eight, tied with the standing end, remains as the outermost rope throughout the knot.

(3) The knot is dressed tightly.

## 7-10. CLOVE HITCH KNOT

The clove hitch knot can be used in the middle of the rope as well as at the end of the rope (Figure 7-10). The knot must have constant tension on it once tied to prevent slipping. It can be used as either an anchor or middle knot, depending on how it is tied. (Do not use a clove hitch knot on metal except on a picket-hold-fast.)

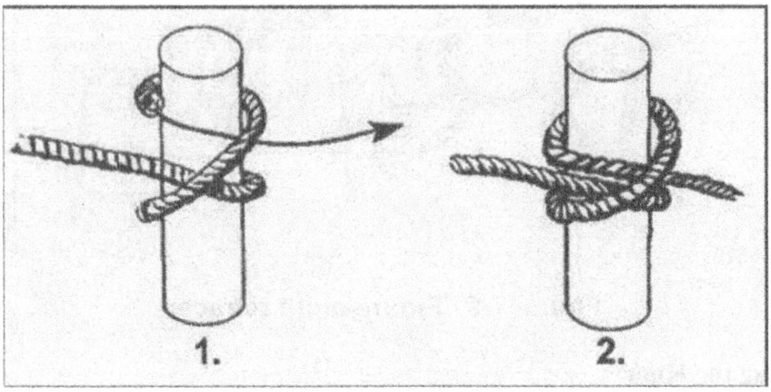


**Figure 7-10. Clove hitch knot.**

a. **Tying the Knot.**

(1) If there is access over the top of the anchor, the knot is tied as follows:

STEP 1. Hold rope in both hands, palms down with hands together. Slide the left hand to the left from 20 to 25 centimeters (8 to 10 inches).

STEP 2. Form a loop away from and back toward the right.

STEP 3. Slide the right hand from 20 to 25 centimeters (8 to 10 inches) to the right. Form a loop inward and back to the left hand.

STEP 4. Place the left loop on top of the right loop. Place both loops over the anchor and pull both ends of the rope in opposite directions to complete the knot.

(2) If the clove hitch knot must go around the anchor, tie the knot as follows (for instructional purposes, assume that the anchor is horizontal):

STEP 1. Place 76 centimeters (30 inches) of rope over the top of the anchor. Hold the standing end in the left hand. With the right hand, reach under the horizontal anchor, grasp the working end, and bring it inward.

STEP 2. Place the working end of the rope over the standing end (to form a loop). Hold the loop in the left hand. Place the working end over the anchor from 20 to 25 centimeters (8 to 10 inches) to the left of the loop.

STEP 3. With the right hand, reach down to the left-hand side of the loop under the anchor. Grasp the working end of the rope and bring it up and outward.

STEP 4. Dress down the knot.

b. **Checkpoints.**
(1) The knot has two round turns around the anchor with locking bar.
(2) The locking bar is facing 90 degrees from the direction of pull.
(3) The ends exit 180 degrees from each other.
(4) The knot has more than a 15-centimeter (6-inch) pigtail remaining.

## SECTION III. MIDDLE ROPE KNOTS

### 7-11. WIREMAN'S KNOT

The wireman's knot forms a single, fixed loop in the middle of the rope (Figure 7-11).

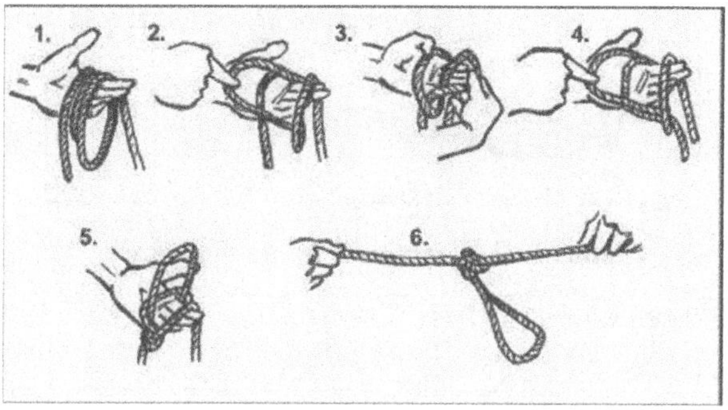


**Figure 7-11. Wireman's knot.**

a. **Tying the Knot.**

STEP 1. When tying this knot, face the anchor that the tie-off system will be tied to. Take up the slack from the anchor, and wrap two turns around the left hand (palm up) from left to right.

STEP 2. Take up a loop of 30 centimeters (12 inches) in the second round turn to create the fixed loop of the knot.

STEP 3. Name the wraps from the palm to the fingertips: heel, palm, and fingertip.

STEP 4. Grasp the palm wrap with the right thumb and forefinger, and place it over the heel wrap.

STEP 5. Grasp the heel wrap and place it over the fingertip wrap.

STEP 6. Grasp the fingertip wrap and place it over the palm wrap.

STEP 7. Grasp the palm wrap and pull up to form a fixed loop.

STEP 8. Dress the knot down by pulling on the fixed loop and the two working ends.

STEP 9. Pull the working ends apart to finish the knot.

b. **Checkpoints.**

(1) The completed knot should have four separate bights locking down on themselves with the fixed loop exiting from the top of the knot and laying toward the near side anchor point.

(2) Both ends should exit opposite each other without any bends.

## 7-12. DIRECTIONAL FIGURE-EIGHT KNOT

The directional figure-eight knot forms a single, fixed loop in the middle of the rope that lays back along the standing part of the rope (Figure 7-12).

**Figure 7-12. Directional figure-eight knot.**

a. **Tying the Knot.**

STEP 1. Face the far side anchor so that when the knot is tied, it lays inward.

STEP 2. Lay the rope from the far side anchor over the left palm. Make one wrap around the palm.

STEP 3. With the wrap thus formed, tie a figure-eight knot around the standing part that leads to the far side anchor.

STEP 4. Dress the knot down with the tail and the bight together.

b. **Checkpoints.**

(1) The loop should be large enough to accept a snaplink but no larger than a helmet-size loop.

(2) The tail and bight must be together.

(3) The figure eight is tied tightly.

(4) The bight in the knot faces back toward the near side.

## 7-13. BOWLINE-ON-A-BIGHT KNOT

The bowline-on-a-bight knot is used to form two fixed loops in the middle of a rope (Figure 7-13).

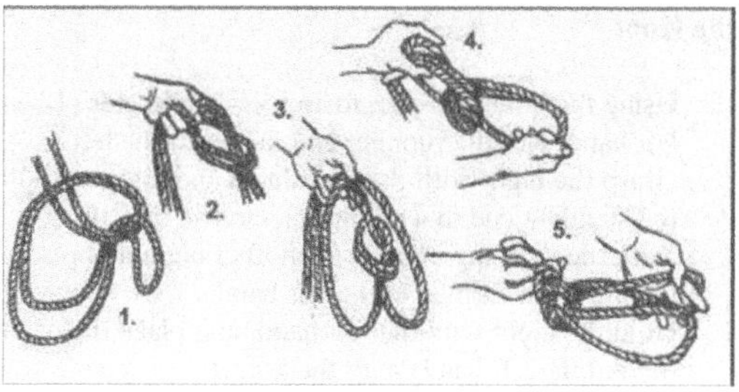


**Figure 7-13. Bowline-on-a-bight knot.**

a. **Tying the Knot.**

STEP 1. Form a bight in the rope about twice as long as the finished loops will be.
STEP 2. Tie an overhand knot on a bight.
STEP 3. Hold the overhand knot in the left hand so that the bight is running down and outward.
STEP 4. Grasp the bight with the right hand, fold it back over the overhand knot so that the overhand knot goes through the bight.
STEP 5. From the end (apex) of the bight, follow the bight back to where it forms the cross in the overhand knot. Grasp the two ropes that run down and outward and pull up, forming two loops.
STEP 6. Pull the two ropes out of the overhand knot and dress the knot down.
STEP 7. A final dress is required: grasp the ends of the two fixed loops and pull, spreading them apart to ensure the loops do not slip.

b. **Checkpoints.**

(1) There are two fixed loops that will not slip.
(2) There are no twists in the knot.
(3) A double loop is held in place by a bight.

## 7-14. FIGURE-EIGHT-ON-A-BIGHT KNOT

The figure-eight-on-a-bight knot is used to form two fixed loops in the middle of a rope . (Figure 7-14).

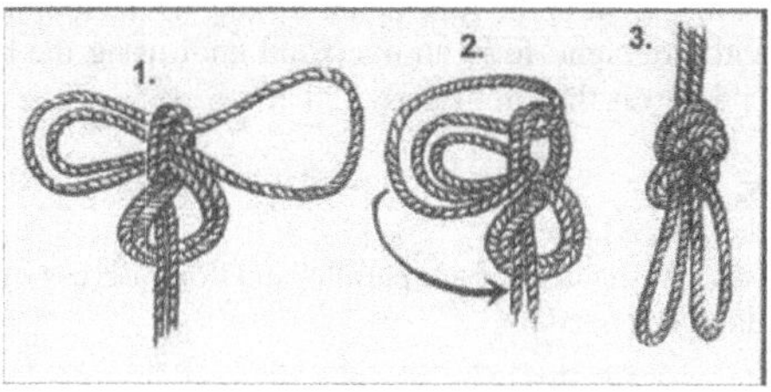


**Figure 7-14. Figure-eight-on-a-bight knot.**

a. **Tying the Knot.**

STEP 1. Using the doubled rope, form a 46-centimeter (18-inch) bight in the left hand with the running end facing to the left.

STEP 2. Grasp the bight with the right hand and make a 360-degree turn around the standing end in a counterclockwise direction.

STEP 3. With the working end, form another bight and place that bight through the loop just formed in the left hand.

STEP 4. Hold the bight with the left hand, and place the original bight (moving toward the left hand) over the knot.

STEP 5. Dress the knot down.

b. **Checkpoints.**

(1) There is a double figure eight with two loops.

(2) The knot will not slip.

## 7-15. OVERHAND LOOP KNOT

The overhand loop knot forms a single loop in the middle of a rope. It should not be used in a transport tightening system (Figure 7-15).

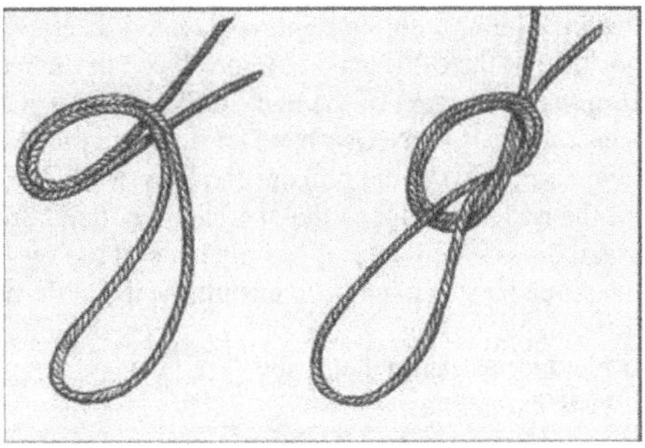

**Figure 7-15. Overhand loop knot**

a. **Tying the Knot.**

STEP 1. Form a bight in the rope about as long as the loop needed.

STEP 2. With the bight, form an overhand knot using the bight as the working end. Dress the knot tightly.

b. **Checkpoints.**

(1) The loop is the desired size.

(2) The ropes in the overhand knot are parallel and do not cross over each other.

(3) The knot is tightly dressed.

## 7-16. FIGURE-EIGHT LOOP KNOT

The figure-eight loop knot is a variation of the overhand loop and is easy to untie after it has held weight (Figure 7-16).

**Figure 7-16. Figure eight loop knot.**

a. **Tying the Knot.**

STEP 1. Form a bight in the rope about as large as the diameter of the desired loop.
STEP 2. With the bight as the working end, form a loop in rope (standing part).
STEP 3. Wrap the working end around the standing part 360 degrees and feed the working end through the loop. Dress the knot tightly.

b. **Checkpoints.**
(1) The loop is the desired size.
(2) The ropes in the loop are parallel and do not cross over each other.
(3) The knot is tightly dressed.

## SECTION IV. SPECIAL KNOTS

## 7-17. SINGLE BUTTERFLY KNOT

The single butterfly knot is used to form a single, fixed loop in the middle of the rope without using the ends (Figure 7-17). The butterfly can be used for the middle man in a rope party as well as in a transport tightening system. The knot can be hard to untie when heavy weight has been placed on it for extended periods.

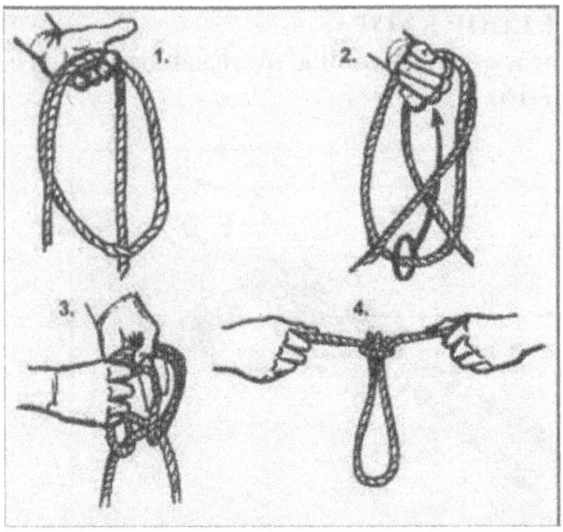


**Figure 7-17. Single butterfly knot.**

a. **Tying the Knot.**

STEP 1. Form a loop over the palm of the left hand.

STEP 2. With the right hand, cross the two long-standing ends; reach under and grasp the bottom of the loop formed in STEP 1.

STEP 3. Route the bottom of the loop over the crossed standing ends and back through the top of the loop.

STEP 4. Hold the newly formed loop and dress the knot down, removing all of the slack out of the knot.

b. **Checkpoints.**

(1) The wings of the knot are dressed down tightly and as close together as possible.

(2) The ropes between the wings are parallel with no crossovers.

(3) The loop is no larger than a helmet but large enough to accept a snaplink.

(4) All ropes in the knot are tightly dressed.

## 7-18. PRUSIK KNOT

The prusik knot is used to put a moveable rope on a fixed rope such as a prusik ascent or a tightening system. This knot can be tied as a middle, finger, or end knot.

a. **Tying the Knot.**

(1) *First method.* The prusik knot is tied at the middle or end of a rope; there are many ways to do this (Figure 7-18). The middle-of-the-rope prusik knot can be tied with a short rope to a long rope as follows:

STEP 1. Double the short rope, forming a bight, with the working ends even. Lay it over the long rope so that the closed end of the bight is 30 centimeters (12 inches) below the long rope and the remaining part of the rope (working ends) is the closest to the rappeller; spread the working end apart.

STEP 2. Reach down through the 30-centimeter (12-inch) bight. Pull up both of the working ends and lay them over the long rope. Repeat this process making sure that the working ends pass in the middle of the first two wraps. Now there are four wraps and a locking bar working across them on the long rope.

STEP 3. Dress the wraps and locking bar down to ensure they are tight and not twisted. Tie an overhand knot with both ropes to prevent the knot from slipping during periods of variable tension.

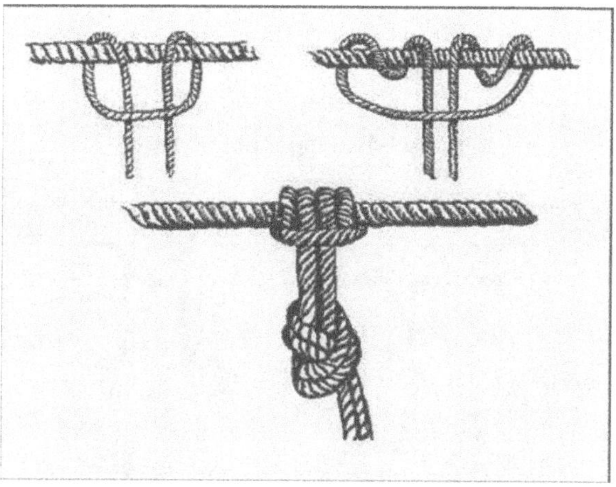

**Figure 7-18. Middle-of-the-rope prusik knot.**

(2) *Second method.* Another way of tying the prusik knot is the finger prusik (Figure 7-19).

STEP 1. Lay the rope across the thumb and index finger of the left hand.

STEP 2. Wrap the rope around the index finger twice (wrapping away from the rappeller). Wrap the rope around the thumb twice (wrapping towards the rappeller).

STEP 3. Place the thumb and index finger tip to tip. Push the wraps together so that the prusik is on the left thumb. Four wraps and a locking bar should be across them.

STEP 4. Pass the running end of the rope through the wraps of the prusik knot.

STEP 5. Dress the wraps and locking bar down to ensure they are tight and not twisted.

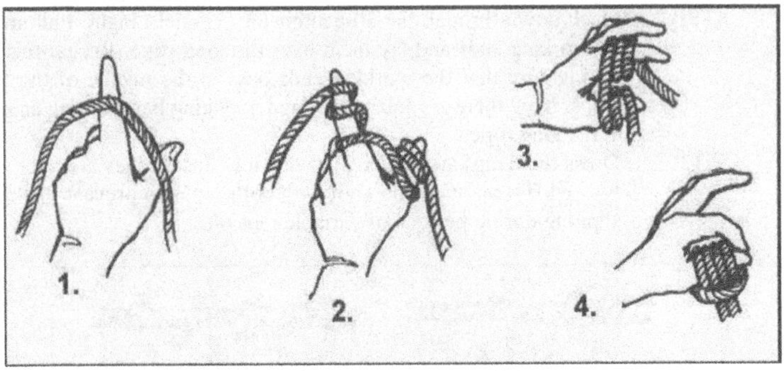


**Figure 7-19. Finger prusik knot.**

(3) *Third method.* Another way of tying a prusik knot is the end-of-the-rope knot (Figure 7-20).

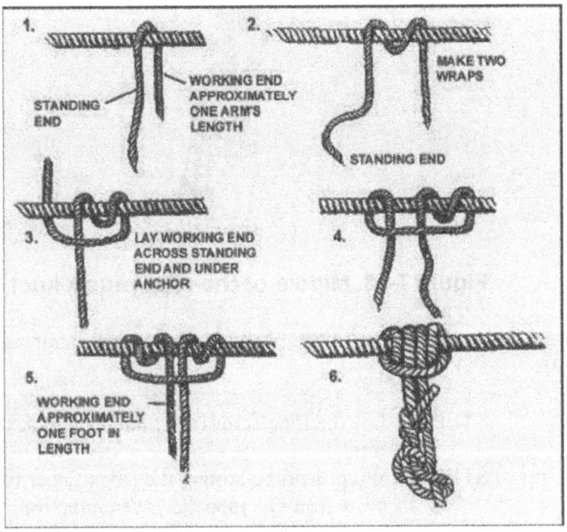


**Figure 7-20. End-of-the-rope prusik knot.**

STEP 1. Using an arm's length of rope, place it over the long rope.
STEP 2. Form a complete round turn in the rope.
STEP 3. Cross over the standing part of the short rope with the working end of the short rope.
STEP 4. Lay the working end under the long rope.
STEP 5. Form a complete round turn in the rope, working back toward the middle of the knot.

STEP 6. Four wraps and a locking bar should run across them on the long rope. Dress the wraps and locking bar down. Ensure they are tight, parallel, and not twisted.

STEP 7. Finish the knot with a bowline to ensure that the prusik knot will not slip out during periods of varying tension.

b. **Checkpoints.**

(1) Four wraps with a locking bar.

(2) The locking bar faces the rappeller.

(3) The knot is tight and dressed down with no ropes twisted or crossed.

(4) Other than a finger prusik, the knot should contain an overhand or bowline to prevent slipping.

## 7-19. BACHMAN KNOT

The bachman knot provides a means of using a makeshift mechanized ascender (Figure 7-21).

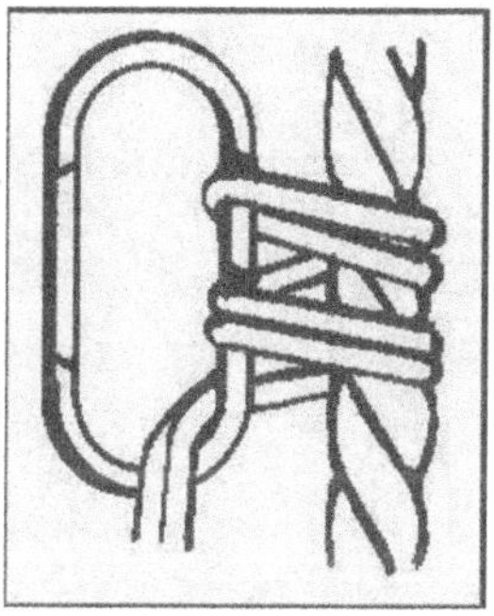

**Figure 7-21. Bachman knot.**

a. **Tying the Knot.**

STEP 1. Find the middle of a utility rope and insert it into a snaplink.

STEP 2. Place the snaplink and utility rope next to a long climbing rope.

STEP 3. With the two ropes parallel fromthe snaplink, make two or more wraps wraps around the climbing rope and through the inside portion of the snaplink.

**Note:** The rope can be tied into an etrier (stirrup) and used as a prusik-friction principal ascender.

b. **Checkpoints.**

(1) The bight of the climbing rope is at the top of the snaplink.

(2) The two ropes run parallel without twisting or crossing.

(3) Two or more wraps are made around the long climbing rope and through the inside portion of the snaplink.

## 7-20. BOWLINE-ON-A-COIL KNOT

The bowline-on-a-coil knot is an expedient tie-in used by rappellers in two- and three-man party climbs when a climbing harness is not available (Figure 7-22).

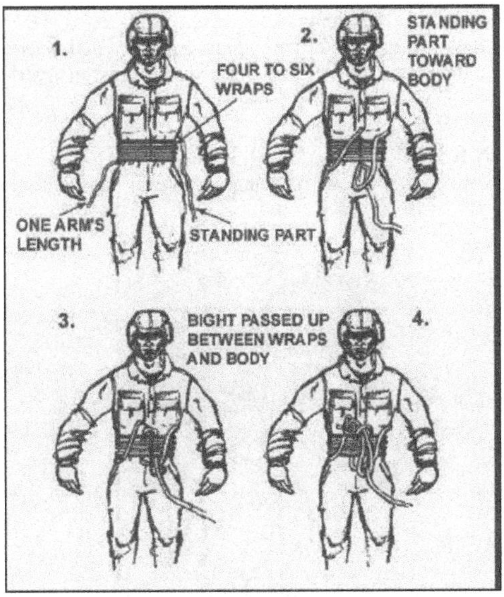


**Figure 7-22. Bowline-on-a-coil knot.**

**Note:** A safer tie in, when possible, is a swiss seat and figure eight around the shoulder, to avoid damaging the ribs if the climber falls.

a. **Tying the Knot.**

STEP 1. To tie a bowline-on-a-coil knot, start the same as tying a bowline around the waist. (The waist of the rappeller is the anchor point. The rappeller has about 40 centimeters [16 inches] of rope in the right hand. The standing end is coming from the remainder of the rope in his left hand.)

STEP 2. Draw slack from the standing end of the rope in the left hand, wrapping enough coils around the body to complete a minimum of four (but no more than six) coils.

STEP 3. With the rope in the left hand, make a loop inward with the standing end.

STEP 4. Slide the loop up between the horizontal wraps and the body, ensuring that the loop does not come uncrossed.

STEP 5. Bring a bight up through the loop. Bring the working end of the rope in the right hand through the bight and back onto itself.

***OR***

Insert the short working end to the right through the loop (just passed through the wraps). Pass the working end around the long standing end and back on itself, forming a bight.

STEP 6. Dress the knot down.

STEP 7. Safety off with an overhand knot against the knot formed on the top single coil. From 10 to 15 centimeters (4 to 6 inches) of rope should be left (standing part).

b. **Checkpoints.**

(1) There are a minimum of four (maximum of six) parallel wraps; the top and bottom ropes cross forward of the hips.

(2) The loop must be underneath all wraps.

(3) Other checkpoints are the same as the bowline.

## 7-21. THREE-LOOP BOWLINE KNOT

The three-loop bowline knot is used to form three fixed loops in the middle of a rope (Figure 7-23). It is used in a self-equalizing anchor system.

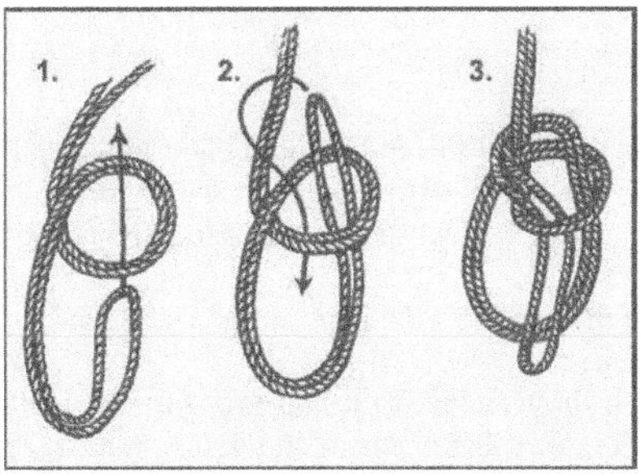


**Figure 7-23. Three-loop bowline knot.**

a. **Tying the Knot.**

STEP 1. Form about a 60 centimeter (24-inch) bight in the rope, laying it across the left hand, palm up, with the bight to the front.

STEP 2. With the right hand, form a loop with both ropes. Hold it in the left hand.

STEP 3. With the right hand, bring the long bight through the loop(formed in STEP 2) and around and around the two standing ends.

STEP 4. Follow the bight around the standing ends and back down into the loop.

STEP 5. Hold all three loops in the left hand. With the right hand, grasp both standing ropes and pull while dressing up the bowline.

b. **Checkpoints.**

(1) Checkpoints are the same as a regular bowline, except that it is tied with two ropes.

(2) An overhand knot must be tied if the third loop is not to be used as a secondary anchor.

## 7-22. FIGURE-EIGHT SLIP KNOT

The figure-eight slip knot forms an adjustable bight in a rope (Figure 7-24).

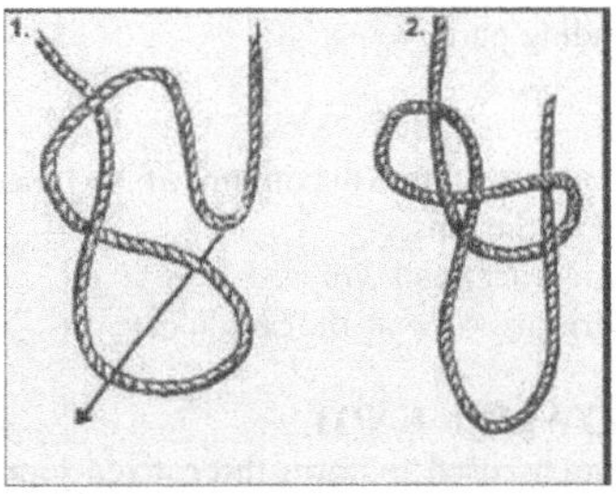


**Figure 7-24. Figure-eight slip knot.**

a. **Tying the Knot.**

STEP 1. Form a 30-centimeter (12-inch) bight in the end of the rope.

STEP 2. Hold the center of the bight in the right hand. Hold the two parallel ropes from the bight in the left hand about 30 centimeters (12 inches) up the rope.

STEP 3. With the center of the bight in the right hand, twist two complete turns clockwise.

STEP 4. Reach through the bight and grasp the long standing end of the rope. Pull another bight (from the long standing end) back through the original bight.

STEP 5. Pull down on the short working end of the rope and dress the knot down.

STEP 6. If the knot is to be used in a transport tightening system, take the working end of the rope and form a half hitch around the loop of the figure-eight knot.

b. **Checkpoints.**

(1) The knot is in the shape of a figure eight.

(2) The sliding portion of the rope is the long working end of the rope.

## Section V. SPECIAL-PURPOSE KNOTS

### 7-23. KLEIMHIEST KNOT

The kleimhiest knot provides a moveable, easily adjustable, high-tension knot capable of holding extremely heavy loads while being pulled tight (Figure 7-25). This system requires a six-man pulling team when used in a transport tightening system.

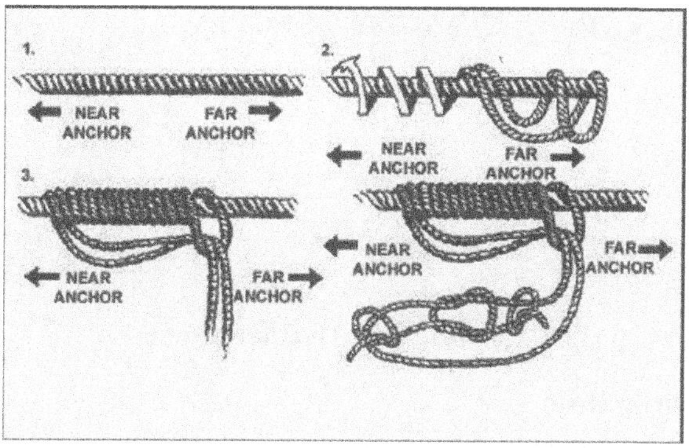


**Figure 7-25. Kleimhiest knot.**

a. **Tying the Knot.**

STEP 1. Using a utility rope, offset the ends by 30 centimeters (12inches). With the ends offset, find the center of the rope and form a bight. Lay the bight over a horizontal rope near the far side anchor.

STEP 2. Wrap the tails of the utility rope around the horizontal rope back toward the near side anchor. Wrap at least four complete turns.

STEP 3. Pass the remaining tails of the utility rope through the bight (see STEP 1).

STEP 4. Join the two ends of the tail with a joining knot (square knot with overhand safety knots).

STEP 5. Dress the knot down tightly so that all wraps are touching. Insert two snaplinks (opposing gates) into the large loop tied with the tails.

b. **Checkpoints.**

(1) The bight is closest to the far side anchor.
(2) All wraps are tight and touching.
(3) The ends of the utility rope are properly secured with the adjoining knot.

## 7-24. OVERHAND KNOT

The overhand knot (Figure 7-26) is used to make a knotted rope for a handline, to secure the ends of other knots, and to make stirrups in direct-aid climbing. This knot can also be used to temporarily whip the end of a rope. It is used with other knots.

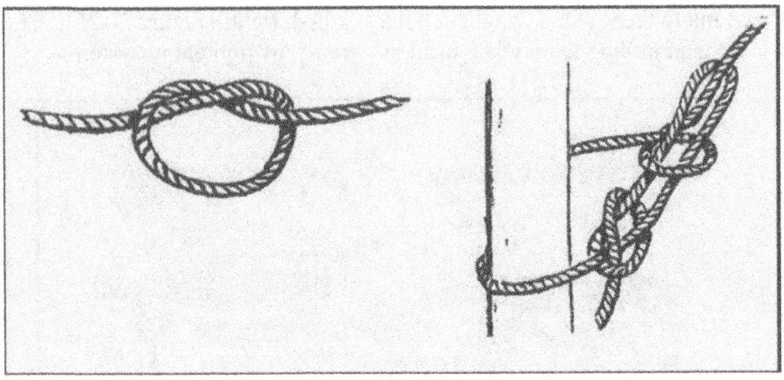

**Figure 7-26. Overhand knot.**

a. **Tying the Knot.**

STEP 1. Make a loop in the rope.
STEP 2. Pass the working end of the rope through the loop.
STEP 3. Dress down the knot.

b. **Checkpoints.**

(1) Running ends exit the loop in opposite directions.
(2) The knot is dressed down tightly.

## 7-25. FROST KNOT

The frost knot is used when working with flat webbing (Figure 7-27). It holds better than a standard overhand knot or square knot.

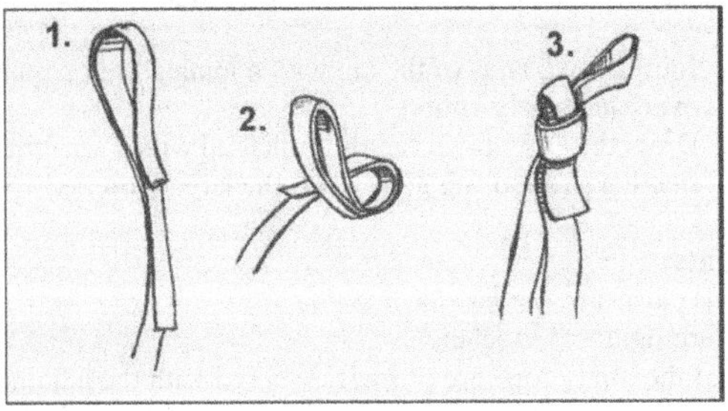


**Figure 7-27. Frost knot.**

a. **Tying the Knot.**

STEP 1. Lap one end (a bight) of webbing over the other (about 25 to 30 centimeters [10 to 12 inches]).

STEP 2. Tie an overhand knot with the newly formed triple-strand webbing; dress tightly.

b. **Checkpoints.**

(1) The tails of the webbing run in opposite directions.
(2) Three strands of webbing are formed into a tight overhand knot.
(3) A bight and tail exit the top of the overhand knot.

## 7-26. GIRTH HITCH KNOT

The girth hitch knot is used in tying a runner to a partly driven piton (Figure 7-28).

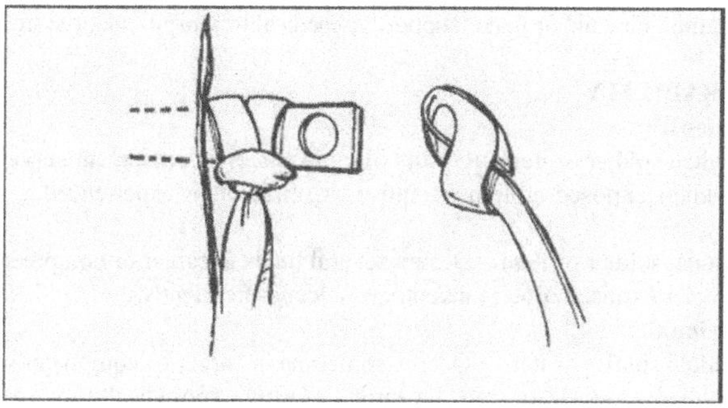

**Figure 7-28. Girth hitch knot.**

a. **Tying the Knot.**

STEP 1. Bring the standing ends back through the bight of a length of rope or webbing.

STEP 2. Dress the knot tightly.

b. **Checkpoints.**

(1) Two wraps exist with a locking bar running across the wraps.
(2) The knot is dressed tightly.

APPENDIX A

# RISK ASSESSMENT

*This appendix explains risk assessment for rappel operations. A risk assessement matrix is shown in Table A-1.*

## A-1. EFFECT

a. **Catastrophic.** Death or permanent total disability, system loss, major property damage.

b. **Critical.** Permanent partial disability, temporary total disability in excess of three months, major system damage, significant property damage.

c. **Moderate.** Minor injury, lost workday accident, compensable injury or illness, minor system damage, minor property damage.

d. **Negligible.** First aid or minor supportive medical treatment, minor system impairment.

## A-2. PROBABILITY

a. **Frequent.**

(1) Individual soldier or item—Occurs often in career or equipment service life.

(2) All soldiers exposed or item inventory—Continuously experienced.

b. **Likely.**

(1) Individual soldier or item—Occurs several times in career or equipment service life.

(2) All soldier exposed or item inventory—Occurs frequently.

c. **Occasional.**

(1) Individual soldier or item—Occurs sometime in career or equipment service life.

(2) All soldiers exposed or item inventory—Occurs sporadically, or several times in inventory service life.

d. **Seldom.**

(1) Individual soldier or item—Possible to occur in career or equipment service life.

(2) All soldiers exposed or item inventory—Remote chance of occurrence; expected to occur sometime in inventory service life.

e. **Unlikely.**

(1) Individual soldier or item—Can assume will not occur in career or equipment service life.

(2) All soldiers exposed or item inventory—Possible, but improbable; occurs only very rarely.

## A-3. RISK LEVELS

a. **Extremely High.** Loss of ability to accomplish mission.

b. **High.** Significantly degrades mission capabilities in terms of required mission standards.

c. **Medium.** Degrades mission capabilities in terms of required mission standards.

d. **Low.** Little or no impact on mission accomplishment.

| RISK ASSESSMENT | | | HAZARD PROBABILITY | | | | |
|---|---|---|---|---|---|---|---|
| | | | Frequ<br>ent | Likel<br>y | Occasio<br>nal | Seld<br>om | Unlike<br>ly |
| | | | A | B | C | D | E |
| EFFECT | Catastrophic | I | ***Extremely High*** | | | | |
| | Critical | II | | | ***High*** | | |
| | Moderate | III | | ***Medium*** | | | |
| | Negligible | IV | | | ***Low*** | | |

**Table A-1. Risk assessment matrix.**

APPENDIX B

# BASIC EQUIPMENT FOR RAPPEL OPERATIONS

*The equipment listed in this appendix is the standard basic items required to conduct ground and air rappelling operations.*

## B-1. GROUND OPERATIONS

For ground operations, the following equipment is needed:

a. Nylon rope, 120 feet (NSN 4020-00-931-8793).
b. Snaplinks (NSN 8465-00-360-0228).
c. Gloves (NSN 8415-00-268-7868).
d. Rope coiling log or standard deployment bag.
e. Rappel seat.
f. Rappel rings (used as expedient anchor points).

## B-2. AIR OPERATIONS

For air operations, the following equipment is needed:

a. All items in paragraph B-1.
b. Donut ring (UH-1H operations only) (locally produced TSC item).
c. Safety floating ring.
d. Two 6-foot safety ropes (with snaplinks).
e. Eight 120-foot nylon ropes.
f. Monkey harness for static rappel master.
g. Sixteen snaplinks (UH-60 operations).
h. Twenty-two snaplinks (UH-1H operations).
i. FM communications.
j. Aircraft communications headset.

# GLOSSARY

## ACRONYMS AND ABBREVIATIONS

| | |
|---|---|
| AFRM | assistant fast-rope master |
| AGL | above ground level |
| AMC | air mission commander |
| AR | Army regulation |
| ATP | aircrew training program |
| attn | attention |
| BDU | battle dress uniform |
| CG | commanding general |
| CGU | modified helicopter safety strap |
| chemlight | chemical light |
| cm | centimeter |
| DA | Department of the Army |
| D-bag | deployment bag |
| F | Fahrenheit |
| FM | frequency modulated; field manual |
| FRIES | fast-rope insertion/extraction system |
| FRM | fast-rope master |
| GSO | ground safety officer |
| HQ | headquarters |
| IAW | in accordance with |
| ICS | interagency communications system |
| intercomm | intercommunication |
| LBE | load-bearing equipment |
| LZ | landing zone |
| MEDEVAC | medical evacuation |
| mm | millimeter |
| mph | miles per hour |
| MSO | mountaineering safety officer |
| NCO | noncommissioned officer |
| NCOIC | noncommissioned officer in charge |
| NSN | national stock number |
| NVG | night vision goggles |

| | |
|---|---|
| NWTC | Northern Warfare Training Center |
| OIC | officer in charge |
| PIC | pilot in command |
| POI | program of instruction |
| POL | petroleum, oils, and lubricants |
| RATELO | radiotelephone operator |
| RSO | rappel safety officer |
| SFC | sergeant first class |
| SOAR | Special Operations Aviation Regiment |
| SOP | standing operating procedures |
| SPIES | special patrol infiltration/exfiltration system |
| STABO | a system for extracting personnel by helicopter (the combined first letters of the surnames of the five persons who designed the system) |
| SSO | STABO system operator |
| TAPO | Technology Applications Program Office |
| TC | training circular |
| TM | technical manual |
| TRADOC | Training Doctrine Command |
| TSC | training support center |
| UIAA | Union of International Alpine Association |
| UH | utility helicopter |
| US | United States |
| USAF | United States Air Force |
| USAIS | United States Army Infantry School |
| USASOC | United States Army Special Operations Command |
| VIP | very important person |

# REFERENCES

## DOCUMENTS NEEDED

These documents must be available to the intended users of this publication.


| | |
|---|---|
| DA Form 2028 | Recommended Changes to Publications and Blank Forms. 1 February 1974. |
| DA Form 5752-R | Rope History and Usage. May 1989. |
| FM 57-38* | Pathfinder Operations. 9 April 1993. |
| FM 7-93* | Long Range Surveillance Unit Operations. 3 October 1995. |
| TC 90-6-1* | Military Mountaineering. 26 April 1989. |


## READINGS RECOMMENDED

These sources contain relevant supplemental information.


| | |
|---|---|
| AR 95-3 | Aviation: General Provisions, Training, and Standardization, A Resource Management. 27 September 1990. |
| FM 8-35 | Evacuation of the Sick and Wounded. 22 December 1983 |
| FM 57-220 | Basic Parachuting Techniques and Training. 1 June 1990. |
| TM 10-1670-262-12 | Operator's and Organizational Maintenance Manual (Including Repair Parts and Special Tools List) for Personnel STABO Extraction System and Cargo Personnel Lowering Anchor Device. 7 April 1988. |
| USASOC Reg 350-2 | Training Airborne Operations. 1 June 1995. |
| USASOC Reg 350-6 | Training Special Operations Forces Infiltration/Exfiltration Techniques. 12 May 1995. |
| | The Air Assault School Handbook. |


*This source was also used to develop this publication.

# INDEX

TC 21-24
10 SEPTEMBER 1997

By Order of the Secretary of the Army:

DENNIS J. REIMER
*General, United States Army*
*Chief of Staff*

Official:

JOEL B. HUDSON
*Administrative Assistant to the*
*Secretary of the Army*
03807

**DISTRIBUTION:**

*Active Army, Army National Guard, and U.S. Army Reserve*: To be distributed in accordance with the initial distribution number 115059, requirements for TC 21-24.